DI

DIGITAL SWITCHING SYSTEMS

OTHER McGRAW-HILL COMMUNICATIONS BOOKS OF INTEREST

Benner *Fibre Channel*
Best *Phase-Locked Loops, Third Edition*
Bush *Private Branch Exchange System and Applications*
Dayton *Integrating Digital Services*
Feit *TCP/IP, Second Edition*
Gallagher *Mobile Telecommunications Networking with IS-41*
Goldberg *Digital Techniques in Frequency Synthesis*
Goralski *Introduction to ATM Networking*
Harte *Cellular and PCS: The Big Picture*
Hebrawi *OSI: Upper Layer Standards and Practices*
Heldman *Information Telecommunications*
Heldman *Competitive Telecommunications*
Kessler *ISDN, Third Edition*
Kessler, Train *Metropolitan Area Networks*
Kurupilla *Wireless PCS*
Lee *Mobile Cellular Telecommunications, Second Edition*
Lee *Mobile Communications Engineering, Second Edition*
Lindberg *Digital Broadband Networks and Services*
Logsdon *Mobile Communication Satellites*
Macario *Cellular Radio, Second Edition*
Rhee *Error Correction Coding Theory*
Roddy *Satellite Communications, Second Edition*
Rohde et al. *Communications Receivers, Second Edition*
Simon et al. *Spread Spectrum Communications Handbook*
Smith *Cellular Design and Optimization*
Tsakalakis *PCS Network Deployment*
Winch *Telecommunication Transmission Systems*

Digital Switching Systems

System Reliability and Analysis

Syed Riffat Ali

Bell Communications Research, Inc.
Piscataway, New Jersey

McGraw-Hill, Inc.
New York • San Francisco • Washington, D.C.
Auckland • Bogotá • Caracas • Lisbon • London
Madrid • Mexico City • Milan • Montreal • New Delhi
San Juan • Singapore • Sydney • Tokyo • Toronto

Library of Congress Cataloging-in-Publication Data

Ali, Syed Riffat.
Digital switching systems : system reliability and analysis / Syed Riffat Ali.
p. cm.
Includes index.
ISBN 0-07-001069-2
1. Telephone switching systems, Electronic—Reliability.
2. Digital telephone systems—Reliability. 3. System analysis.
I. Title.
TK6397.A483 1997 97-19408
621.387—dc21 CIP

McGraw-Hill

A Division of The ***McGraw-Hill*** *Companies*

1 2 3 4 5 6 7 8 9 0 FGR/FGR 9 0 2 1 0 9 8 7

ISBN 0-07-001069-2

The sponsoring editor for this book was Steve Chapman, the editing supervisor was Scott Amerman, and the production supervisor was Pamela Pelton. It was set in Vendome ICG by Tanya Howden and Jennifer L. Dougherty of McGraw-Hill's Professional Book Group composition unit, Hightstown, N.J.

Printed and bound by Quebecor/Fairfield.

McGraw-Hill books are available at special quantity discounts to use as premiums and sales promotions, or for use in corporate training programs. For more information, please write to the Director of Special Sales, McGraw-Hill, 11 West 19th Street, New York, NY 10011. Or contact your local bookstore.

This book is printed on recycled, acid-free paper containing a minimum of 50% recycled, de-inked fiber.

In memory of my father,
Syed Mohammed Taqi Razavi,
who dedicated his life to the pursuit of
better education and happiness for his children.

TRADEMARK ACKNOWLEDGEMENTS

CONTENTS

PREFACE

The motive of this book is to expose practicing telephone engineers and other graduate engineers to the art of digital switching system (DSS) analysis. The concept of applying system analysis techniques to the digital switching systems as discussed in this book evolved during the divestiture period of the Bell Operating Companies (BOCs) from AT&T. Bell Communications Research, Inc. (Bellcore), formed in 1984 as a research and engineering company supporting the BOCs, now known as the seven Regional Bell Operating Companies (RBOCs), conducted analysis of digital switching system products to ascertain compatibility with the network. Since then Bellcore has evolved into a global provider of communications software, engineering, and consulting services.

The author has primarily depended on his field experience in writing this book and has extensively used engineering and various symposium publications and advice from many subject matter experts at Bellcore.

This book is divided into six basic categories. Chapters 1, 2, 3, and 4 cover digital switching system hardware, and Chaps. 5 and 6 cover software architectures and their impact on switching system reliability. Chapter 7 primarily covers field aspects of digital switching systems, i.e., system testing and acceptance and system maintenance and support. Chapter 8 covers networked aspects of the digital switching system, including STP, SCP, and AIN. Chapter 9 develops a hypothetical digital switching system and explores system analysis needs. The object of Chap. 10 is to introduce the hardware and software of various major digital switching systems in use today in the North American network.

Summary

This book develops hardware and software architectures related to modern digital switching systems. It uses with an inside-out approach, i.e., starting from basic building blocks and then gradually adding more advanced functionalities. It then applies advanced techniques to hardware and software reliability analysis for class 5 (end-office) applications.

This book covers Markov chain analysis for hardware reliability and an enhanced root-cause and metrics model methodology for improving software reliability.

The first four chapters introduce various aspects of digital switching system architecture, the internal communications and control necessary for digital switching system subsystems, and the basics of switching fabric, i.e., switching elements within a switch. Chapters 3 and 4 develop basic theory related to reliability modeling based on Markov chain analysis. Chapters 5 and 6 introduce the basics of digital switching system software architecture and software analysis methodologies. Chapter 7 discusses some practical aspects of digital switch maintenance. Chapter 8 extends Markov chain analysis to networked switching elements such as the signaling transfer point (STP). Chapter 9 develops a generic digital switching system and discusses switching system analysis needs. Chapter 10 describes some basic hardware and software architectures of major digital switching systems currently deployed in the North American Network.

Major subjects covered under each chapter are as follows:

Chapter 1: Switching System Fundamentals: Central office linkages, switching system hierarchy, building blocks of digital switching system, and basic call processing.

Chapter 2: Communications and Control: Levels of control, basic functionalities of digital switch subsystems, control architectures, multiplexed highways, switching fabric, programmable junctors, network redundancy.

Chapter 3: Reliability Modeling: Downtimes in digital switching systems, reliability assessment techniques, Markov models, failure models, and sensitivity analysis.

Chapter 4: Switching System Reliability Analysis: Analysis of central processor community, clock subsystem, network controller subsystem, switching network, line and trunk downtimes, call cutoffs, ineffective machine attempts, and partial downtimes.

Chapter 5: Digital Switching Software: Software architecture, operating systems, database management, digital switching system software classification, call models, software linkages during a call, and feature interaction.

Chapter 6: Quality Analysis of Switching System Software: Life cycle of switching system software, software development, a methodology for assessing switching software quality, software testing, CMM, and other models.

Chapter 7: Maintenance of Digital Switching Systems: Interfaces of central office, system outage causes, software patches, generic program upgrade, problem reporting, firmware deployment, maintainability metrics, a strategy for improving software quality, and defect analysis.

Chapter 8: Analysis of Networked Switching Systems: Switching in a networked environment, network reliability requirements, Markov model of a hypothetical STP, dependence of new technologies on digital switch, ISDN, AIN, and future trends in digital switching systems.

Chapter 9: A Generic Digital Switching System Model: Hardware architectures of a central processor, network control processors, interface controllers, interface modules, software architecture, recovery strategy, calls through the switching system, and analysis report.

Chapter 10: Major Digital Switching Systems in the North American Network: High-level hardware and software architectures.

Acknowledgments

The author extends his appreciation and gratitude to Gobin Ganguly, Fred Hawley, Chi-Ming Chen, Adrian Dolinsky, and Bob Thien for their help and support.

DIGITAL SWITCHING SYSTEMS

CHAPTER 1

Switching System Fundamentals

1.1 Introduction

This book assumes that the reader is familiar with the fundamentals of telephony. The author will use concepts and language commonly known and understood in the telecommunications industry. However, for the reader who needs to review some basic concepts, a reference list is provided at the end of this chapter.[1]

A telephony system can be divided into three categories:

1. Circuit switching
2. Station equipment
3. Transmission

This book will deal primarily with practical, as opposed to theoretical, aspects of the first category above, circuit switching. References will be made to the other two categories, station equipment and transmission systems, only when necessary for the reader to understand the functions of circuit switching; otherwise, these subjects will not be covered in this book.

1.2 Digital Switching System Analysis

System analysis and design is defined as *the process of developing user requirements and designing systems to achieve them effectively.*[2] The main object of this book will be to expose the reader to various aspects of digital switching system requirements, design analysis, and mathematical techniques which could help him or her to analyze a digital switching system.

An *exchange,* or *central office,** is a large, complex system comprising many subsystems, each with unique characteristics and functionality. A basic understanding of these subsystems and their interaction with the rest of the system is needed for a digital switching system to be effectively analyzed.

---

* Throughout this book, the term central office is used instead of exchange. The term exchange is predominantly used in Europe and Asia.

1.2.1 Purpose of Analysis

The reliability of digital switching systems is becoming increasingly important for users of telephone services. Currently, most Internet access takes place through digital switching systems. Almost all electronic money transfers depend on the reliability of digital switching systems. The federal government requires that all network outages exceeding 30 minutes be reported to the Federal Communications Commission (FCC). The Bell Operating Companies require that the outage of digital switches not exceed 3 minutes per system per year. All this shows that the reliability of digital switching systems is a very serious matter; it can impact a nation's commerce, security, and efficiency. But how does the engineering profession ensure that the telecommunications network is reliable and will operate within prescribed limits? This, in effect, is what this book tries to answer. There are many excellent books on telecommunications, networks, and reliability. However, this book attempts to bring these disciplines under an umbrella of reliability analysis.

Headlines like "String of Phone Failures Perplexes Companies and U.S. Investigators"[3] or "Regional Phone Systems on Both Coasts Are Disrupted by Glitches in Software"[4] appear frequently in the news media. The nature of such system outages needs to be understood and methodologies put in place that will alleviate such problems. The current telephone network is becoming very complex; it has multiple owners and is equipped by many different suppliers. This book will address the root cause of many outages, which usually involves a breakdown of a network element such as a digital switching system.

Digital switching systems represent very complex systems. They are multifaceted and require an analyst to explore many avenues during the analysis process. This book provides guidance and answers to some core questions that an analyst would usually ask. The following questions and answers should give you an idea of how this book goes about that task:

Question: How does one start to analyze the reliability of a complex product such as a digital switching system?

Answer: By fully understanding the workings of a digital switching system before attempting to access its reliability.

Question: What types of reliability data does one need to analyze a digital switching system?

Answer: In short, the answer varies with the extent of the reliability analysis one needs to conduct. If one needs to conduct only the hardware reliability assessment of a digital switching system, then data pertaining to hardware components, such as failure rates, repair times, and hardware architecture, are needed. If the analysis also includes software assessment, then the process requires an understanding of software design methodologies, software architecture, software quality control, software testing, etc.

Question: Can one apply generic techniques in conducting reliability analysis of a digital switching system?

Answer: Yes. Application of such generic techniques is the primary objective of this book. The methodology is flexible enough to be applied not only to digital switching systems but also to other network elements, with some modifications.

Question: There are so many different types of digital switching systems in the world. How can one begin to understand them?

Answer: A comprehensive understanding of the various digital switching systems is another objective of this book. The reader will first be introduced to the hardware and software architectures of a hypothetical generic digital switching system. These concepts can then be extended to any commercially available digital switching system or to new systems under development.

The following approach is taken in this book to establish a basis for the analysis of digital switching systems:

- To better understand the architecture of a digital switching system, a hypothetical generic digital switching system is developed. This model has all the high-level subsystems usually found in commercial digital systems.
- The path of some common calls through the generic digital switching system is traced, explicating simple call flow through commercial digital switching systems.
- Communications and control that are normally required for digital switching systems are described.
- How a digital switching system uses different types of call switching technologies is explained.
- Reliability models that best describe different subsystems of a typical digital switching system are explored.
- The software architecture of digital switching systems along with assessment and prediction of software quality are also covered.

- Operational and maintenance issues of a digital switching system that may impact its operational reliability are explored.
- Reliability models for network elements that interface with digital switching systems are created.

With this roadmap, we seek to better understand digital switching systems and develop methodologies that could be used to assess their reliability.

1.2.2 Basic Central Office Linkages

During the analysis of a digital switching system, it is helpful to define the extent of a central office (CO)* and its linkages to other facilities. Figure 1.1 shows a typical central office along with some important facilities. Familiarity with this setup is essential to better understand various operations that may impact the overall reliability of a digital switching system.

The following relate to the basic linkages of a typical central office:

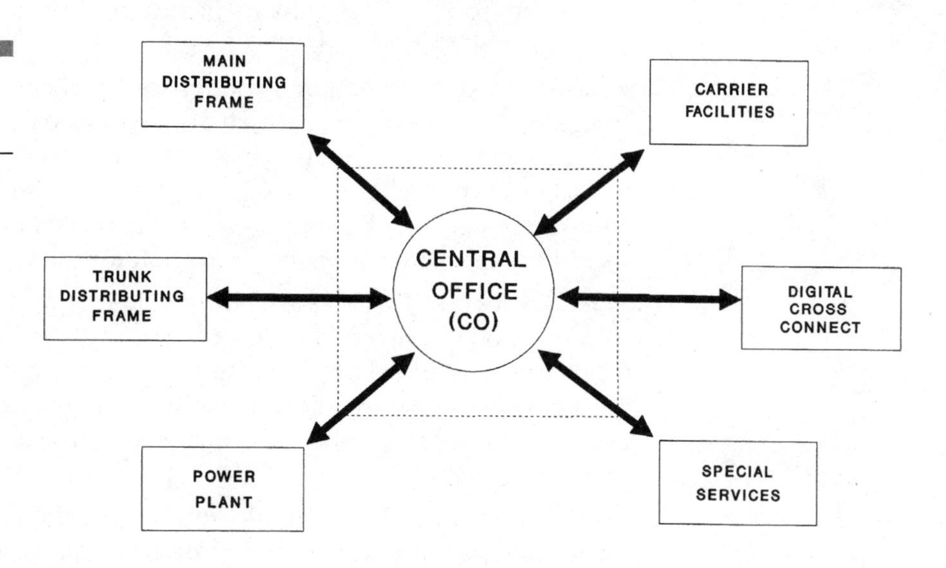

Figure 1.1
Basic central office linkages

* The word "CO" or "switch" for Central Office will be used interchangeably throughout this book.

Main distributing frame (MDF) Location where all lines and other related links are cross-connected to a central office switch, also referred to as the *line side* of a switch. The MDF is probably the most labor-extensive part of a CO. All lines from subscribers terminate in the MDF. The MDF has two sides: a vertical and a horizontal. The subscriber cables terminate on the vertical side. The wiring from the digital switching system referred to as *line equipment* terminates on the horizontal side. Based on the assignment of subscribers to line equipment, wires are connected between the vertical (cable pair) and the horizontal (line equipment pair). The assignment process for subscribers to line equipment is usually automated.

Trunk distributing frame (TDF) Location where all trunks and other related links are cross-connected to a central office switch, also referred to as the *trunk side* of a switch. The TDF is usually smaller than the MDF. All trunk cabling from different locations terminates in the TDF. The TDF has two sides: a vertical and a horizontal. The trunk cables terminate on the vertical side. The wiring from the digital switching system, referred to as *trunk equipment,* terminates on the horizontal side. Based on the assignment of cable to trunk equipment, the vertical cable pair are connected to the horizontal trunk equipment pair. The assignment process for trunks to trunk equipment is usually automated.

Power plant A combination of power converters, battery systems, and emergency power sources which supply the basic −48- and +24-V direct-current (dc) power and protected alternating-current (ac) power to a CO switch or a group of switches. These should not be confused with the power distributing frames in the central offices that provide special voltage conversions and protection for the CO.

Carrier facilities Facilities which provide carrier or multiplex transmission mode between central offices and with other parts of the telephony network. These facilities typically employ coaxial cables (land or undersea) and radio and satellite systems. The carrier facilities usually terminate on the TDF for cross-connection to the digital switching system.

Digital X-connect Digital cross-connect provides automatic assignments and cross-connection of trunks to digital switching systems. It can be considered a small switching system for trunks.

Special services Those services which require special interfaces or procedures to connect central office facilities to a customer, e.g., data services and wireless services.

These terms are defined early in this book so that the reader may clearly understand the basic linkages that drive a central office.

1.2.3 Outside Plant versus Inside Plant

Most of the telephone companies classify their telephone equipment as *outside plant* or *inside plant.* This classification becomes important during the analysis of a switching system, since indirectly it defines the extent of a CO and consequently the scope of analysis. As shown in Fig. 1.1 and explained above, any element of telephony equipment outside the CO box, such as MDF and carrier systems, is classified as *outside plant.* CO equipment, such as central processors, switching fabric, and tone generators, are considered *inside plant.*

1.3 Switching System Hierarchy

Calls through the North American network follow a hierarchical path. The search for a path through the network for a long-distance call follows a hierarchy similar to that in Fig. 1.2.[5] After a call leaves a class 5 switch, a path is hunted through the class 4 office followed by class 3, class 2, and class 1. In addition, there are international gateway offices (extension of class 1) which a central office calls to complete international-destination calls through cables, satellite, or microwaves. Figure 1.2 also shows the different classes of switching system in the North American network:

- *Local exchange (class 5).* It is also referred to as the *end office (EO).* It interfaces with subscribers directly and connects to toll centers via trunks. It records subscriber billing information.
- *Tandem and toll office (class 4).* Most class 5 COs interface with the tandem offices. The tandem offices primarily switch trunk traffic between class 5 offices; they also interface with higher-level toll offices. Toll operator services can be provided by these offices.
- *Primary toll center (class 3).* The class 3 toll center can be directly served by class 4 or class 5 offices, depending upon the trunk deployment. In other words, if the normal number of trunks in these offices are exhausted, then traffic from lower-hierarchy offices can home into a class 3 office. Class 3 offices have the capability of storing, modifying, prefixing, translating, or code-converting

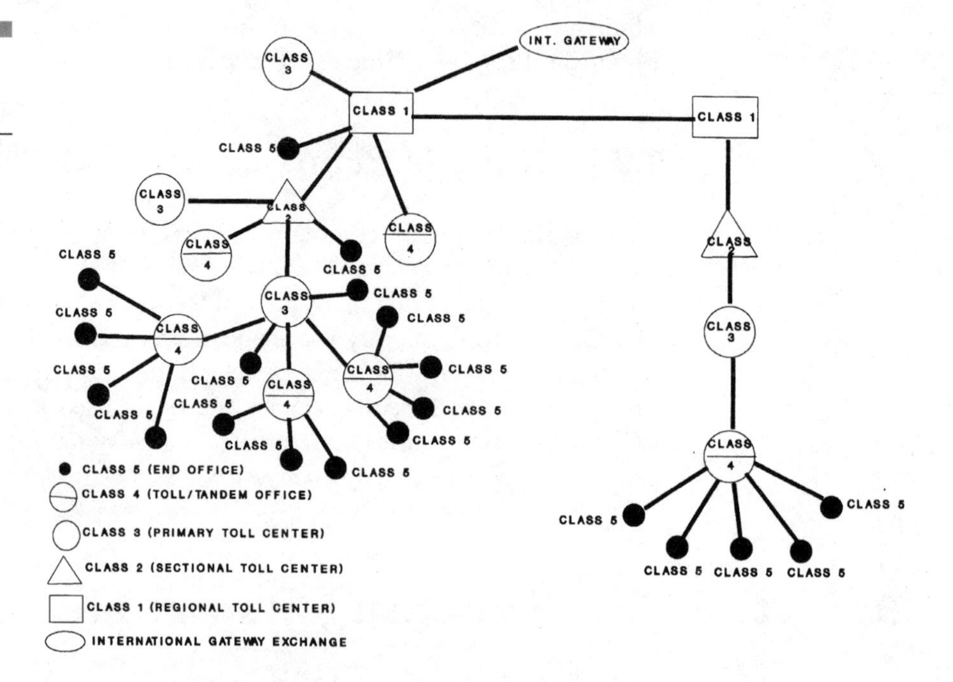

Figure 1.2
Switching system hierarchy

received digits as well as finding the most efficient routing to higher-level toll offices.

- *Sectional toll center (class 2).* It functions as a toll center and can home into class 1 offices.
- *Regional toll center (class 1).* It functions as a toll center and can home into international gateway offices.
- *International gateway.* These offices have direct access to international gateway offices in other countries. They also provide international operator assistance.

The advantage of the hierarchical network is that it provides an efficient way of searching for a path through the network. The disadvantage is that if the primary, sectional, or regional toll center goes down, then large areas of North America can become inaccessible. There are schemes in which some alternate routes are made available, but they cannot carry the full-service load. With the advent of super toll switchers that can switch large numbers of trunks, the number of toll centers is dwindling, which makes the overall network more vulnerable to regional communication blackouts. The following table shows approximate numbers of end offices and toll centers in North America.[6]

Class	Type	Number in 1977	Number in 1982
5	End office	19,000	19,000+
4	Toll and tandem office	1,300	925
3	Primary toll center	230	168
2	Section toll center	67	52
1	Regional toll center	12	10

These figures show that the number of class 5 COs is increasing while the number of toll centers is decreasing.

This switching hierarchy and the classification of offices are covered here to emphasize that just analyzing the reliability of a digital switching system may not solve the problem of overall network reliability. However, to the analyst who understands the interconnection of digital switching systems, it is clear that every part of a switching network must be analyzed to fully appreciate the impact of network reliability. Many class 5 COs also have class 4 capabilities. And most of the class 1, 2, and 3 offices are variants of class 4 architecture. The main objective of this book is to establish analytical techniques that are directly applicable to class 5 COs, and many of these techniques can be extended to include other network elements.

1.4 Evolution of Digital Switching Systems

The next few sections discuss the evolution of digital switching systems as background to understand the current architecture of modern digital switching systems. Many questions about the design rationale for current digital switching systems can be answered by looking at its history. Many design concepts come from the electromechanical telephony switching systems of the past. For instance, the control structure, call handling, alternate routing, billing, etc., all evolved from earlier crossbar switching systems. In fact, the very early electronic switching system used modified crossbar switches as its switching matrix, which we refer to as *switching fabric* in this book.

1.4.1 Stored Program Control Switching Systems

With the advent of software-controlled central processors, the control of switching functions was programmed into memory and actions were executed by the controlling processor. The early versions of electronic switching systems had temporary memory for storing transient-call information and semipermanent memories that carried programming information and could be updated. A *stored program control* (*SPC*) switching system, shown in Fig. 1.3, depicts a simplified view of a telephony switch.

The basic function of an SPC system is to control line originations and terminations and to provide trunk routing to other central or tandem offices. The SPC system also provided control of special features and functions of a central office, identified here as *ancillary control.* The intelligence of an SPC system resided in one processor, and all peripherals were controlled by this single processor. These processors were duplicated for reliability. A modern digital switching system employs a number of processors and uses distributed software and hardware architectures. These functions are developed and explained in later chapters of this book. Control of the maintenance functions of the modern digital switching system also evolved from earlier SPC systems. These systems depended heavily on a single processor to conduct all maintenance functions of the switch. Most of the modern digital switching systems employ a separate processor for maintenance functions. The mainte-

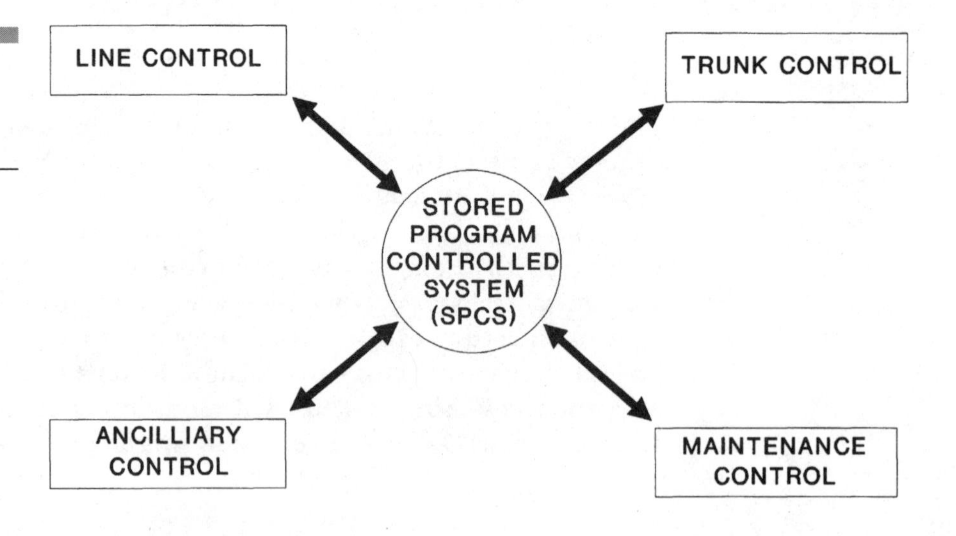

Figure 1.3 Basic control structure of a central office

nance functions of a digital switching system are so important that Chap. 7 is devoted to the explanation of various functionalities of a central office maintenance subsystem.

1.4.2 Digital Switching System Fundamentals

Now we extend the basic concept of SPC switching systems to modern digital switching systems. Many basic elements of the digital switching system already exist in the SPC switching system. A switching system is called *digital* when the input to and output from the switching system network can directly support digital signals. A *digital signal* can be defined as coded pulses that can be used for signaling and control. However, analog signals can still be processed through the digital switching system via analog-to-digital (A/D) or digital-to-analog (D/A) converters. This presents a very simplistic view of a digital switching system. However, eventually we develop this simple digital switching system concept into a more elaborate model.

The evolution to digital switching from analog switching is shown in Figure 1.4*a* to *d*. Figure 14*a* shows a typical analog switch with analog lines and trunks. This figure also shows the line side and trunk side of a switch. This separation of the line and trunk functions of a switch is seen throughout the book. Although this figure shows only lines and trunks for simplicity, it is understood that other types of inputs and outputs to a switching system are not shown. As mentioned earlier, the basic function of a switching system is to switch lines and trunks. Many other advanced switching functions are handled by digital switching systems. However, the main objective of digital switching systems is to switch subscribers and trunk facilities. Figure 1.4*b* shows the next step in the evolution of digital switching. This phase uses analog lines and analog trunks but employs A/D and D/A converters for digital processing of calls. The switching element in this illustration is, however, a digital switch, which means that digitized signals are sent through the switch. Figure 1.4*c* shows the next step in this process, in which digital switches can "talk" to other digital switches via digital trunks while simultaneously supporting analog lines and trunks. Figure 1.4*d* shows the ultimate, an all-digital linkage. In this arrangement, there are no analog lines or trunks involved; all communication between digital switches is via digital signaling. This plan assumes that all incoming lines coming to a CO are digital and that all outgoing trunks are digital as well.

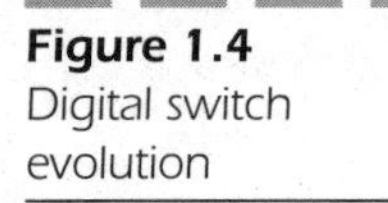

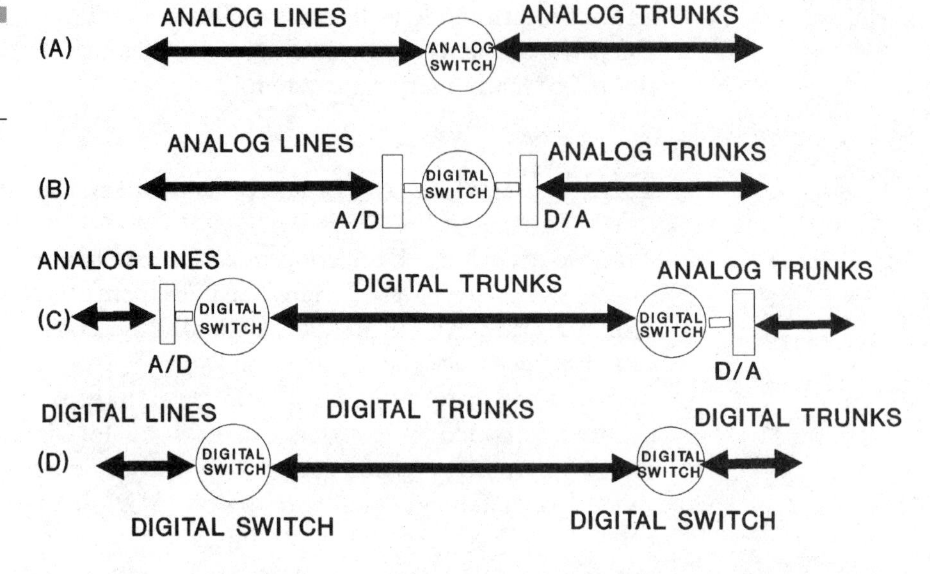

Figure 1.4 Digital switch evolution

In some applications, switching from digital trunk to digital trunk is indeed performed. But we still live in an analog world, and lots of conversion from analog to digital signals is performed in many applications. Currently, the telecommunications industry is moving in the direction in which video, audio, and telephony services will all be combined and switched through digital switching systems. Obviously this will require lots of conversions. One form of switching not shown in Fig. 1.4 is *optical switching*. In this author's view, the future of telephony switching will be optically based. Optical switching systems will provide high-speed, large-bandwidth switching. Currently, many of the "pure" optical switches are under development, and many advances have been made in this area. In the case of optical switching, electrical/optical (E/O) and optical/electrical (O/E) conversions will be required. Fiber-optic–based trunks and lines will be utilized, and signals with very wide bandwidths will be switched. For further information on optical switching, see References at the end of this chapter.

1.4.3 Building Blocks of a Digital Switching System

This book takes an "inside-out" approach to understanding and developing various concepts associated with a modern digital switching system.

Since the object of this book is to systematically analyze digital switching systems, first a basic digital switching system model is developed, and then is expanded gradually throughout the book to cover most of the important functions associated with a modern digital switching system. The development of this digital switching system model is described in four stages. The first stage looks at the very basic kernel of a digital switching system, with the switching matrix, which is called *switching fabric* in this book, since not all switching systems use a matrix arrangement for switching. The switching fabric switches lines and the trunks under the control of a central processor and network controller. The second stage of this development introduces the concepts of line and trunk modules. The third stage introduces the notion of interface controllers and distributed processing. The fourth stage presents a high-level design of a digital switching system equipped with service circuits.

Stage 1 Stage 1 of conceiving a digital switching system is shown in Fig. 1.5*a*. At this stage, all inputs and outputs to a digital switching system are defined. In this particular case, which is a simple one, only lines and trunks are defined. Clearly, there can be many types of lines and trunks. Their design specification must be defined and their reliability assessed at this stage (covered later in this book). The line and trunk sides of the digital switching system are shown separately. As mentioned earlier, this is only a convention and does not mean that trunks appear on one side of the network and lines on the other. The central processor controls the network controller, which in turn controls the switching fabric. For the time being, regard the switching fabric as a "switched" path through the CO. In later chapters we focus on various types of switching fabric architecture with reliability modeling of central processors and network controllers.

Stage 2 Stage 2 of digital switching system design is shown in Fig. 1.5*b*. The concept of line modules (LMs) and trunk modules (TMs) is introduced here. The line and trunk modules are the building blocks of a modern digital switching system, and conceptually they represent some lines or trunks grouped together on circuit packs, termed *line* or *trunk equipment*, and connected to the switching fabric through a controlling interface. Modern digital switching systems use various schemes to terminate lines on the line module. Some digital switching systems allow termination of only one line on one line module, while others allow termination of multiple lines on a single line module. Both schemes have pros

Figure 1.5a
First stage with lines and trunks

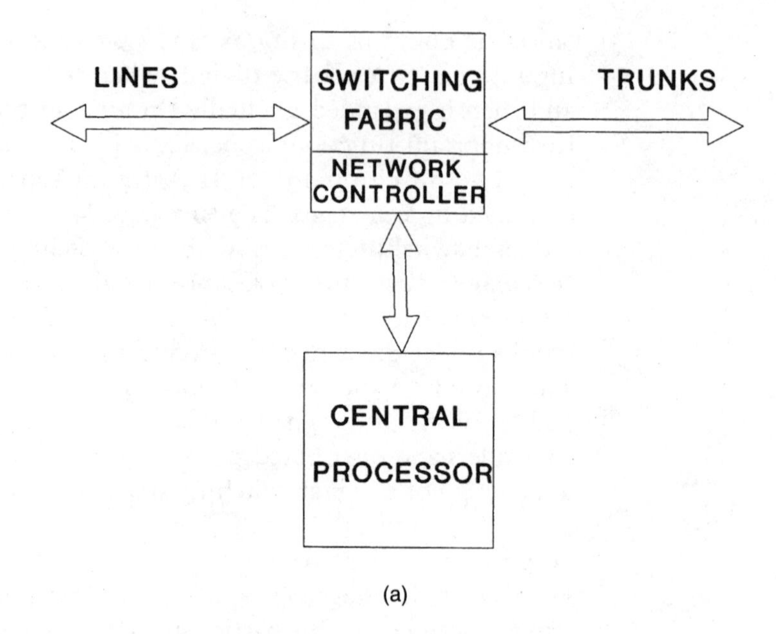

Figure 1.5b
Second stage with line modules and trunk modules

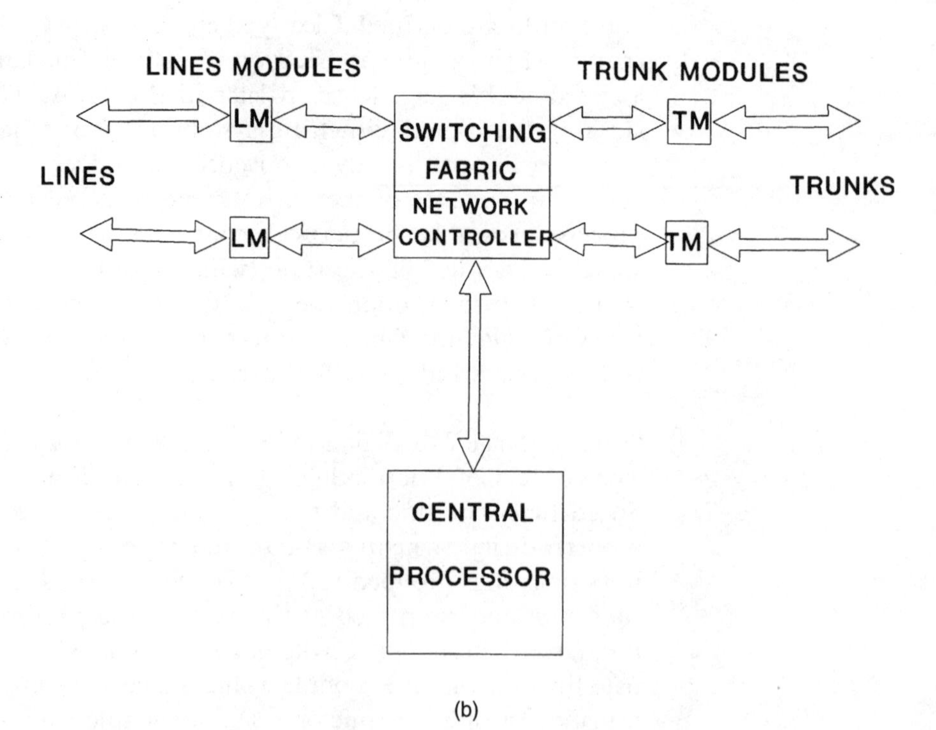

and cons. If a line module becomes defective, this may impact a number of lines if the line module carries multiple lines. However, if a piece of line equipment becomes defective, the line can easily be assigned new line equipment if the LM carried multiple pieces of line equipment. Similar schemes are used for trunks on trunk modules. In a modern digital switching system, line and trunk modules are designed to be modular, which in simple terms means that a number of these units can be added on an as-needed basis without reengineering the system. This allows for easy growth and offers flexibility in offering new services. The impact of these design ideas on system reliability and on digital switching system operation is explored more fully in later chapters.

Stage 3 This stage is depicted in Fig. 1.5*c*. The concept of distributed processing in a digital switching system environment is developed here. Notice the replacement of the network controller in Fig. 1.5*b* by network control processors in Fig. 1.5*c* and the addition of an interface controller for LMs and TMs. The task of controlling the switching fabric is usually assigned to a series of network control processors that control a part of the switching fabric and a group of LMs and TMs. The central processor controls the actions of the network control processors. This type of architecture is very flexible and allows the construction of different sizes of central office by increasing the number of network control processors. For instance, a small central office could be constructed by using just one network control processor, while a configuration of several network control processors could be employed for a larger central office. Naturally, the processing capacity of the network control processors and of the central control processor and network size also play an important role in determining the ultimate size of a central office. This type of architecture is used by many commercial digital switching systems. Later we show how this type of architecture can be modeled and its reliability assessed.

Stage 4 As shown in Fig. 1.5*d*, stage 4 of the digital switching system design may appear to be the final stage of a digital switching system model, but it is not. In reality, it is only an initial model of a digital switching system which is needed to develop a more detailed model. This basic model introduces the *duplicated* scheme now commonly used in modern digital switching systems. Since telephony processing is a nonstop process requiring high reliability, a duplicated

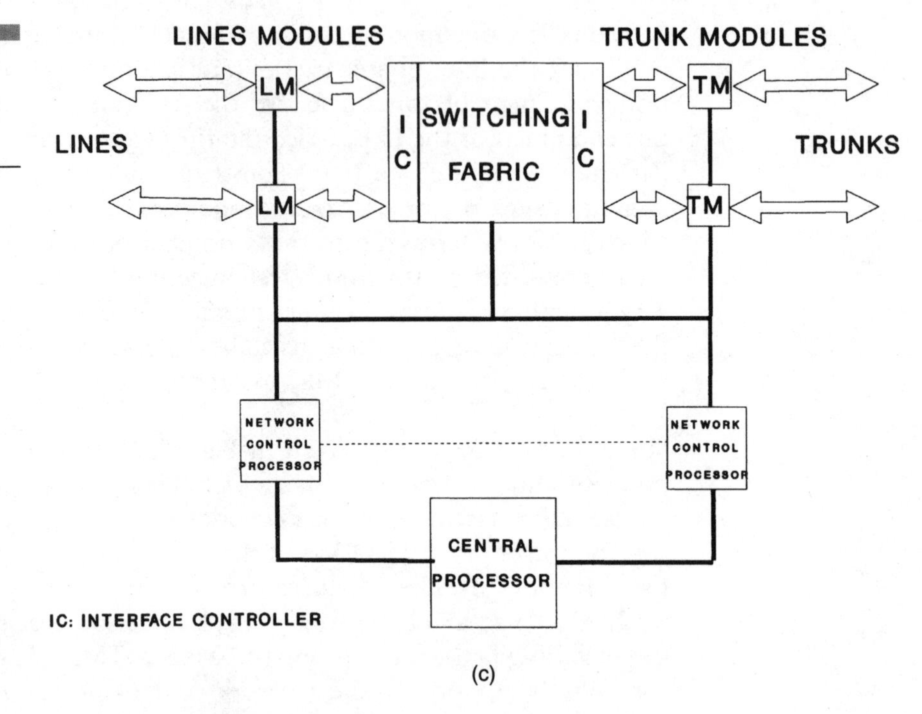

Figure 1.5c Third stage with network control processors

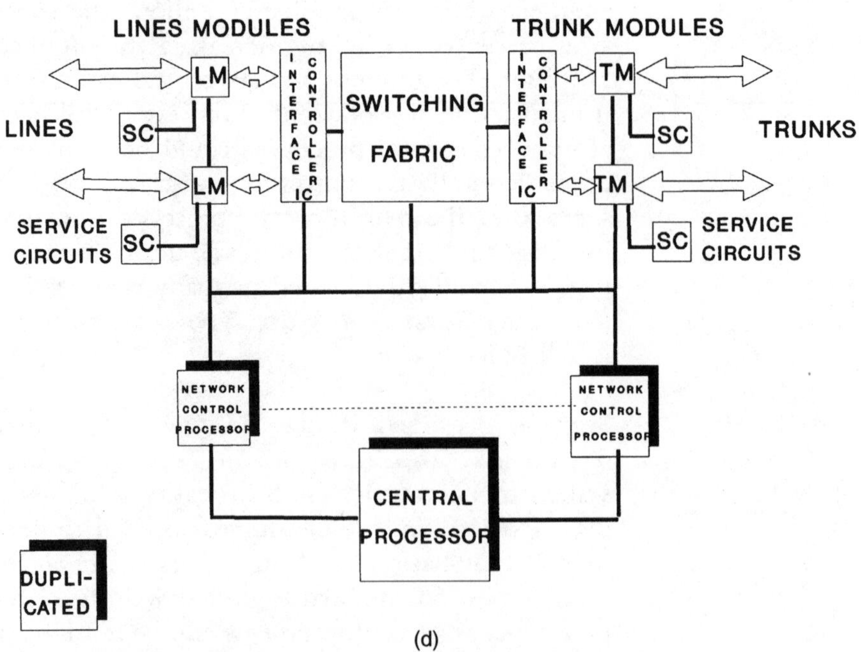

Figure 1.5d Fourth stage with redundant processors

scheme for processing units and associated memory units is almost mandatory. We discuss various duplication schemes and their impact on digital switching system reliability in later chapters. This basic conceptual model also shows the attachment of interface controllers and service circuits to the line and trunk modules. The interface controllers allow interfacing and control of LMs and TMs through the network control processors. The purpose of the service circuits is to provide dial tone, ringing, and other associated functions. In a modern digital switching system, each line or trunk module or a group of modules can be attached to service circuits.

1.4.4 Basic Call Processing

This section describes some *basic* types of calls that are usually processed through a digital switching system:

- Intra-LM calls
- Inter-LM calls
- Incoming calls
- Outgoing calls

As emphasized before, these are basic call classifications and do not reflect any enhanced features or call types. Basic call processing can be easily outlined by a simple digital switching system model, as shown in Fig. 1.5*d*. Knowledge of how calls flow through a digital switching system will make clear the advantages of reliability modeling.

Intra-LM Calls When a customer dials from a telephone that is connected to a specific line module and calls another customer who is also connected to the same line module, this type of call is classified as an intra-LM call. A call path for this type of call is shown in Fig. 1.6*a*. The off-hook (line origination request) condition is detected by the line module, and service circuits are attached to supply a dial tone to the calling customer. Many other functions are performed before a dial tone is given to a calling customer; these are discussed in later chapters. The line module's request for a path through the switching fabric is processed by the interface controller, which in turn works with the network control processor to make a path assignment. Consequently, a path is established through the

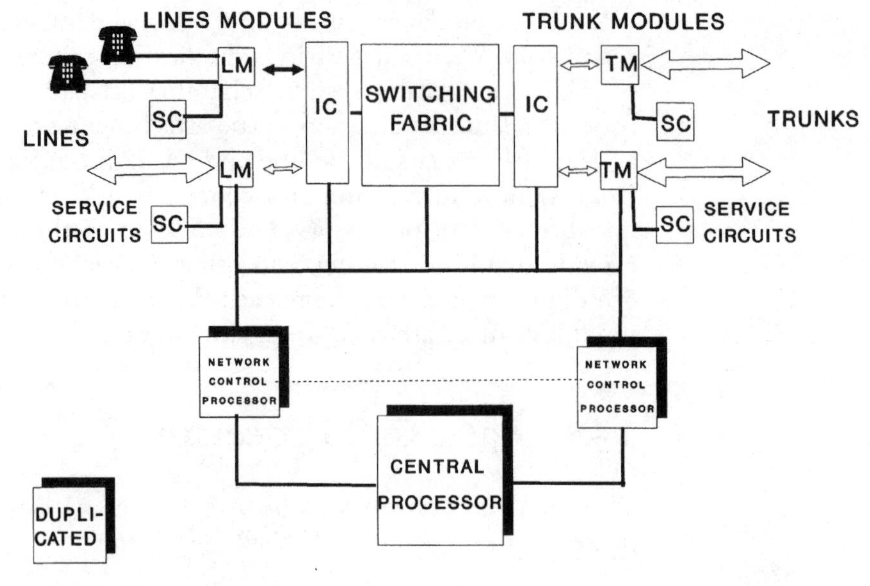

Figure 1.6a
Calls within a line module

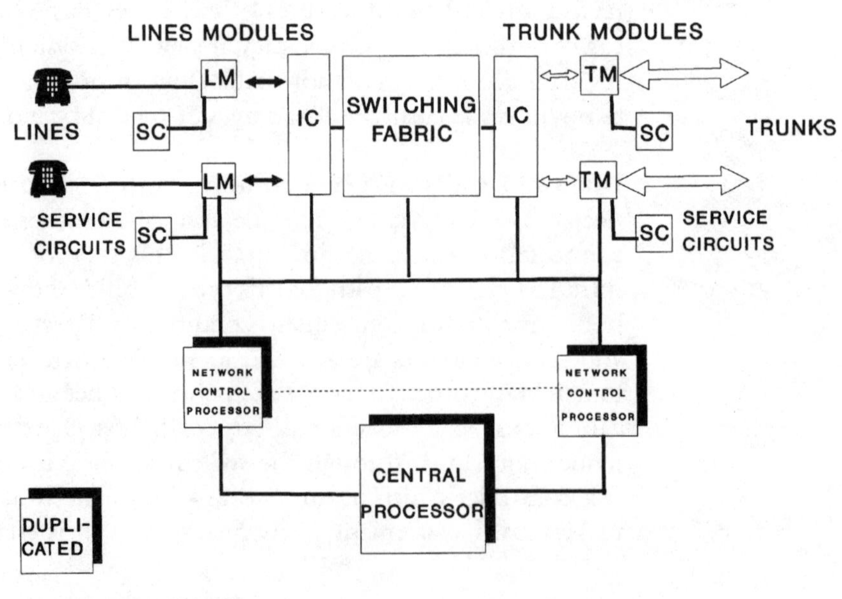

Figure 1.6b
Calls outside a line module

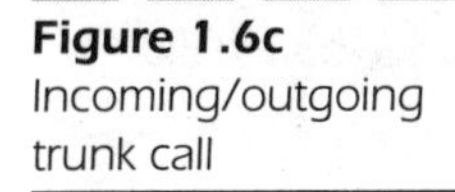

Figure 1.6c Incoming/outgoing trunk call

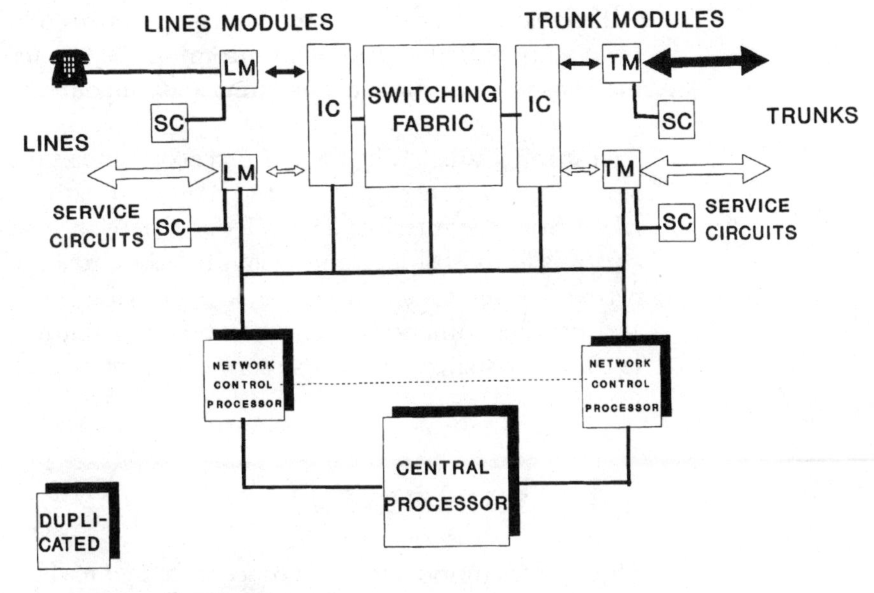

switching fabric for the called line, and a service circuit is attached to ring the line. Again, many other functions are performed before ringing is applied to the called customer; these are also discussed later. Since this is an intra-LM call, the same line module will be involved in controlling the origination and termination of a call. This very simplified explanation is offered here for introductory purposes only. Later chapters go into far greater detail in explaining various functions such as digit reception, digit translation, and tests that are performed before a call is completed.

Inter-LM Calls The workings of an inter-LM call are similar to those of an intra-LM call, except that the terminating line equipment is located in another line module. Figure 1.6*b* shows interconnections for such a call. There are some subtle differences in how an inter-LM call is handled versus an intra-LM call, which are discussed in later chapters.

Outgoing Calls When a LM processes a call which has terminating equipment outside the CO, the LM requests a path through the switching fabric to a trunk module via the interface controller. The interface controller works with the network control processor to establish a path to an outgoing trunk. Once a path is established through the switching fabric, the TM connects a service circuit for controlling the call to the called CO or a tandem office. Functions such as outpulsing and multifrequency

(MF) signaling are provided by the trunk service circuits. An outgoing call from an originating office is an incoming call to a terminating office. Figure 1.6*c* shows the paths of incoming and outgoing calls.

Incoming Calls When a TM detects an incoming call, it attaches service circuits to control the call and requests a path through the switching fabric from the interface controller and network control processor. Once a path is found through the switching fabric to a LM that has the terminating line, service circuits are attached to ring the called telephone. This also provides functions such as audible ringing to the calling line. Use Fig. 1.6*c* to visualize this simple connection of an incoming call.

1.5 Summary

This chapter introduced basic concepts related to digital switching systems, basic CO linkages, the switching hierarchy, and the various evolutionary stages in the development of a digital switching system. Some basic calls were traced to explicate the digital switching system model.

REFERENCES

1. J. C. McDonald, *Fundamentals of Digital Switching*, Plenum Press, New York, 1990.
2. *Encyclopedia of Computer Science and Engineering*, Van Nostrand Reinhold, New York, 1983, p. 729.
3. E. L. Andrews, *The New York Times*, July 2, 1991.
4. M. L. Carnevale, *The Wall Street Journal*, June 27, 1991.
5. Telephone Communications Systems, *Direct Distance Dialing and Toll Systems*, ed. R. F. Rey, vol. 4, Western Electric, 1970, p. 1.13.
6. Bell Laboratories, *Engineering and Operations in the Bell System*, AT&T Bell Laboratories, Murray Hill, NJ, 1977 and 1984, p. 109.

CHAPTER 2

Communications and Control

2.1 Introduction

Chapter 1 introduced the "inside-out" approach to understanding some basic functions of modern digital switching systems. This chapter expands each functional block of the digital switching system discussed so far and treats them as subsystems. These subsystems are studied from functional deployment and architectural points of view.

2.1.1 Scope

The block diagram developed in Chap. 1 of a hypothetical digital switching system reflects the characteristics of many operational switching systems. Here we expand upon the block diagram to better understand the communication and control process of digital switching systems. The subsystems of digital switching systems are broadly classified by their functions and are studied in greater detail in this and the next few chapters. The high-level functionalities can be classified into the following seven categories:

1. Switching communication and control
2. Switching fabric
3. Central processing units
4. Network control processing units
5. Interface controllers
6. Line and trunk circuits
7. Service circuits and central office signaling

2.2 Switching Communication and Control

In this section we discuss communication and control of this hypothetical digital switching system. Figure 2.1 categorizes control of the digital switching system into three levels. These levels are arbitrary and have been introduced to illustrate the control structures of a digital switching system.

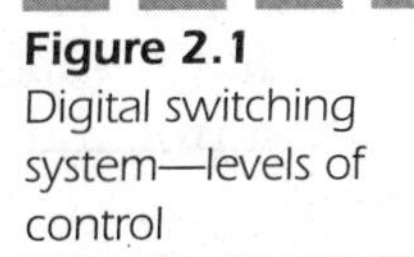

Figure 2.1
Digital switching system—levels of control

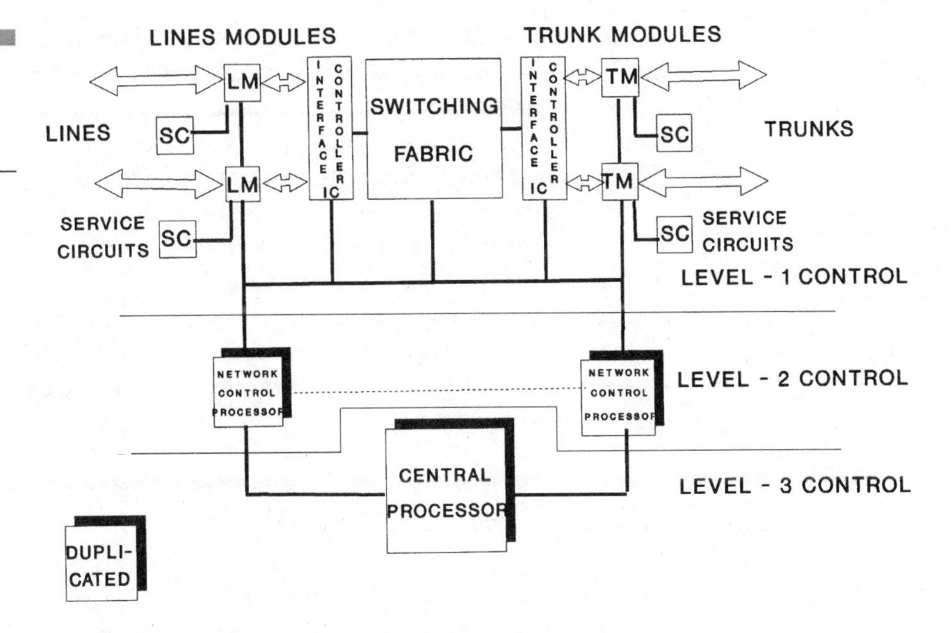

2.2.1 Level 1 Control

Level 1 is the lowest level of control. This level is usually associated with lines, trunks, or other low-level peripherals. Software involved at this level is mostly firmware-based.[1] However, some switching systems do use random access memory (RAM) based software as well. The basic function of software modules at this level is to provide line and trunk access to the interface controllers and low-level call processing support. Other types of call processing support are covered in later chapters. Special control of some features may also be implemented at this level. Most of the line and trunk modules are microprocessor-based and communicate with interface controllers via messages. The function of the interface controllers is to interpret these messages, take certain actions, and in turn communicate with the network control processors. The line and trunk modules are the first interface points at which incoming lines and trunks seek service from a digital switching system. As mentioned earlier, subscriber lines are connected to the main distributing frame (MDF) via cables. A hard-wired cross-connection is established between the subscriber cable pair and the digital switching system's line equipment. Most central offices have a protective device between the subscriber line and the MDF, and these devices are usually installed on protection frames. All lines are continuously scanned by line-scanning programs, which usually reside in the line module. When a customer goes off-hook, the line-scanning program detects this off-hook condition and reports it to the interface controller. All outgoing trunk control and incoming trunk connection requests are handled at this level of control. The interface controller is the primary peripheral controller, and it controls all peripherals associated with call or trunk processing. Any advanced features that a digital switching system needs to support may require special peripheral control modules and would be put under the control of an interface controller. The interface controller in turn is controlled by the network control processor. At this stage only various levels of control are being defined; more details on exactly what needs to be controlled are given later.

2.2.2 Level 2 Control

Level 2 or the midlevel control is usually associated with network controllers and associated functions. These functions are dependent on digital switching system architecture and could reside at this level of control. In a distributed environment, most of the digital switching systems employ mini-size processing units for this level. This is the most important level of control for distributed processing. Some digital switching systems employ a number of network control processors at this level of control. A dedicated bus system is usually required for the processors to communicate with one another. A message format is established for interprocessor communications. Most digital switching systems use messaging protocols for communication between processors. For messaging between the peripherals and external systems, many digital switching systems utilize standard protocols such as Signaling System 7 (SS7), X.25, and X.75. The network control processors are duplicated for reliability. In later chapters we discuss exactly how the duplication process improves the reliability of a subsystem.

2.2.3 Level 3 Control

Level 3 is usually associated with the central processor of a digital switching system. Normally at this level the digital switching systems employ mainframe-type computers. All basic controls of a digital switching system are incorporated at this level. Most of the maintenance and recovery functions of a switch are also controlled from this level. The messaging protocols between the central processor, network control processors, and the switching fabric are established in the design architecture. Private buses between these subsystems are sometimes employed for diagnostic and system recovery purposes.

2.2.4 Basic Functions of Interface Controller

The architecture of this hypothetical digital switching system requires that the interface controller act as an intermediary control element between the line modules and the network control processors. When the customer attached to a line module goes off-hook, the line module detects the off-hook condition and asks the interface controller for validation of the subscriber's line with basic information such as its class of service, subscribed features, and any restrictions on the line. The interface controller maintains a database of subscriber information and validates the line for service. Once the line is validated, the interface controller attaches a service circuit to the line, and a dial tone is provided to the subscriber. After the customer dials, the digits are forwarded to the network control processor with a request for a path through the switching fabric. The network control processor sends a message to the central processor with a translation request of dialed digits. More details on call processing are given in later chapters. Here we introduce some basic messages that need to be processed during simple call handling.

2.2.5 Basic Functions of Network Control Processor

As mentioned previously, more detailed subscriber information is usually stored in the network control processors (NCPs). The NCP also tracks call paths for each call it establishes. During call processing the NCP requests the central processor to translate dialed digits. Once it receives the translated information, that is, the destination of the call, the network control processor hunts for a path through the switching fabric. After the path is established, the NCP keeps track of the call and idles the path, once the call is disconnected.

2.2.6 Basic Functions of Central Processor

As discussed earlier, the central processing system is defined as having level 3 control for this particular architecture, which means it has access to all subsystems of the digital switch. This access could be direct or indirect (i.e., via other subsystems) according to the type of control required. Although the basic functions of a central processing unit are dependent on the switching system architecture, the basic functions of most central processors are essentially the same. Some architectures give more autonomy to certain subsystems than others. Functions that usually require assistance from the central processor are shown in Fig. 2.2, and can broadly be divided into the following categories:

- Call processing
- Network control
- Signaling control
- Maintenance and administration

2.2.7 Call Processing

At the highest level, as shown in Fig. 2.2, basic call processing functions for a central processor usually consist of

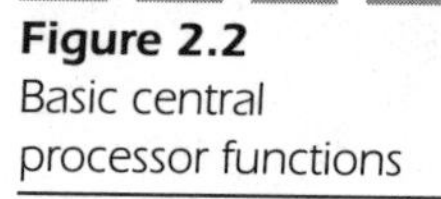

Figure 2.2
Basic central processor functions

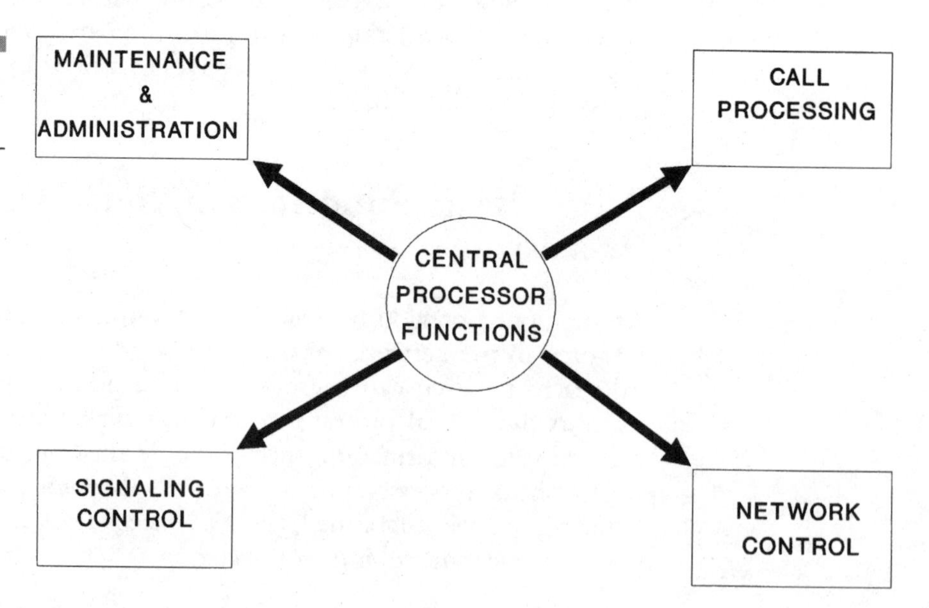

- Incoming/outgoing call translation
- Digit translation
- Call routing and call status
- Billing

In this section we provide a brief synopsis of call processing functions that are usually controlled by a central processor. In a modern digital switching system, a large portion of call processing is performed by the line modules and interface controllers. However, since most digital switching systems still use a central processor, some types of call processing tasks are assigned to it. Most tasks that central processors execute can be termed *global*-level tasks, meaning that the outcome affects all other processors. The distributed nature of current digital switching systems requires that at least one processor in the system keep track of all functions that are common to other processors, especially during call processing or local subsystem recovery. Sometimes, a network control processor can be dedicated to keep track of other processors or to conduct local system recovery. However, most digital switching systems use a central processor for the purposes mentioned above.

Call Translation One important aspect of call processing is the translation of incoming or outgoing digits from a central office to the actual route the call would take. The term *translation* used here pertains to the software lookup tables in the central or network control processor memories. These tables are used to interpret dialed digits and establish operational attributes associated with incoming and outgoing calls, such as type of service and allowed features. Proper routing of calls through the switching system, management of call status, billing, and other functions require network control processor coordination. Tasks are usually assigned to the central processor, since it is the central point for information processing that is common to all network control processors. Each network control processor has information on all subscribers assigned to its interface controllers, but it has no information on the assignment of subscribers to other network control processors. However, it is possible to have an architecture in which the network control processors are kept updated on an assignment of all subscribers, but this would require larger memory for the network control processors and a mechanism to keep the information in all the network control processors always updated. Usually the path setup information for each call is maintained by the central processor since the processor needs a global picture of all call paths through the switching fabric. Once a request comes from the network control processor for a path through the switching fabric, the central processor hunts for a path and keeps it reserved for intended call connection. This path depends on the type of call made. If the subscriber makes a call within the same central office, then the path sought is for a line module in the same central office. If the translation shows that the call is for another central office, then a path is searched for an outgoing trunk to a central office or a tandem office. Different call scenarios may require different paths that need to be hunted through the switching fabric. Detailed call processing schemes and call models are covered in later chapters.

2.2.8 Control Architectures

Here we describe different types of control architectures that many digital switching systems have used in the past or are currently using. The control architecture of a digital switching system usually reflects the extent of control exercised by the central processor. An architecture of a digital switching system may be entirely based on direct controls from a central processor, or the switching system may not employ any central processor at all. Most modern digital switching systems employ an architecture which falls somewhere between these two approaches. Based on this criterion, digital switching systems can be classified into the following four types of control architecture:

- Centralized control
- Hierarchical control
- Quasi-distributed control
- Distributed control

Centralized Control-Based Architecture This type of control architecture is shown in Fig. 2.3. This architecture depends solely on the central processor to control all functions of a digital switching system. The basic functions controlled by the central processor are as follows:

Management of Call Status In this architecture, the line module directly interfaces with the central processor for call processing and call management. When a customer goes off-hook, the line module informs the cen-

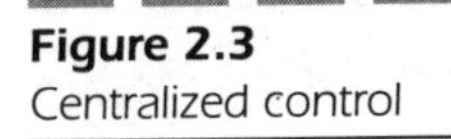

Figure 2.3
Centralized control

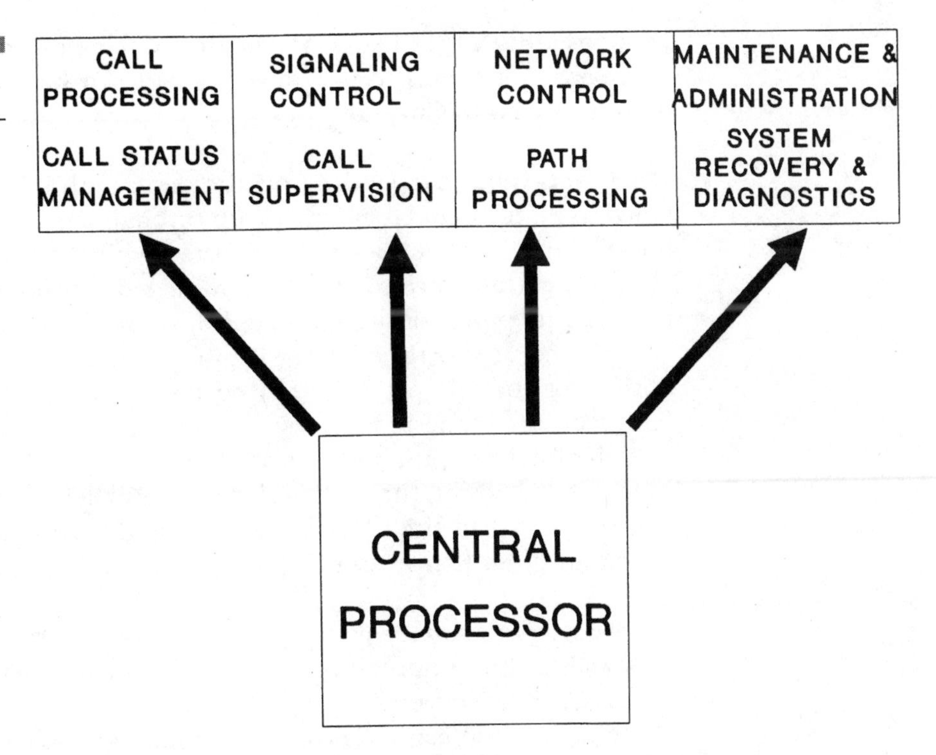

tral processor that a customer has gone off-hook and needs a dial tone. The central processor searches its database for information pertaining to this customer and validates the customer's line. Once the line is validated, the central processor attaches a service circuit to the customer and supplies a dial tone. After the customer dials the number, the central processor analyzes the dialed number and hunts for a path through the switching fabric to establish the call. After the call is established, the central processor establishes status control of the call by marking different bits in its memory as on or off, sometimes referred to as *busy/idle bits.* The busy/idle bits that are assigned to various states of a call are kept current in the central processor memory and are updated during various stages of a call.

Call Supervision Once a call path is established, in the centralized control architecture the linkages associated with a call, such as ringing, connection, disconnection, and signaling, are recorded in the central processor's memory and controlled by the central processor. Regardless of the architecture, if a call is made between two central offices, the responsibility for call supervision always belong to the *originating central office*. Software linkages during a call can be quite complex, and this is covered in Chap. 5.

Path Processing Depending on the design, different techniques are used to hunt for and establish a path through the switching system fabric. Control and the memory associated with these paths are retained by the central processor in the centralized-control-type architecture. The central processor with this type of architecture keeps an "image" of all idle and used paths of the switching fabric. Based on the search algorithm for the idle path, the central processor reserves a path for call connections requested by the line or trunk modules. Under some search algorithms, multiple paths are searched, and the most efficient one is selected for the connection. Most switching fabrics switch in space rather than time. Details concerning different types of switching fabric are given later in this chapter.

System Recovery and Diagnostics The process of isolating faults (diagnostics) and finding a minimum working configuration (system recovery), is discussed fully in later chapters. However, under the centralized control architecture, the central processor handles all diagnostic and system recovery functions along with call processing. For the central processor to be effective, it must have sufficient processing power; otherwise, the switching system's call-carrying capacity and grade of service will be impacted.

Hierarchical Control-Based Architecture The hierarchical control architecture is shown in Fig. 2.4. In this architecture the central processor dedicates some of the tasks of the digital switching system to a number of network control processors. Since a hierarchy of NCPs is added and different tasks are assigned to the NCPs, this type of control architecture can be classified as hierarchical. In this architecture the central processor controls the NCPs, call processing, and the management of call status for all calls in the switch. The NCPs are usually mini-size processors while the central processor is the mainframe type. In this type of architecture, the central processor carries less load than the central processor with a centralized control architecture. Since the NCPs share the load of the central processor, the call-carrying capacity of digital switching systems using this architecture is usually enhanced. The basic drawback of this architecture is the lack of complete modu-

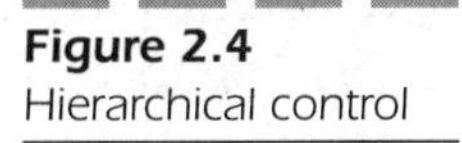

Figure 2.4
Hierarchical control

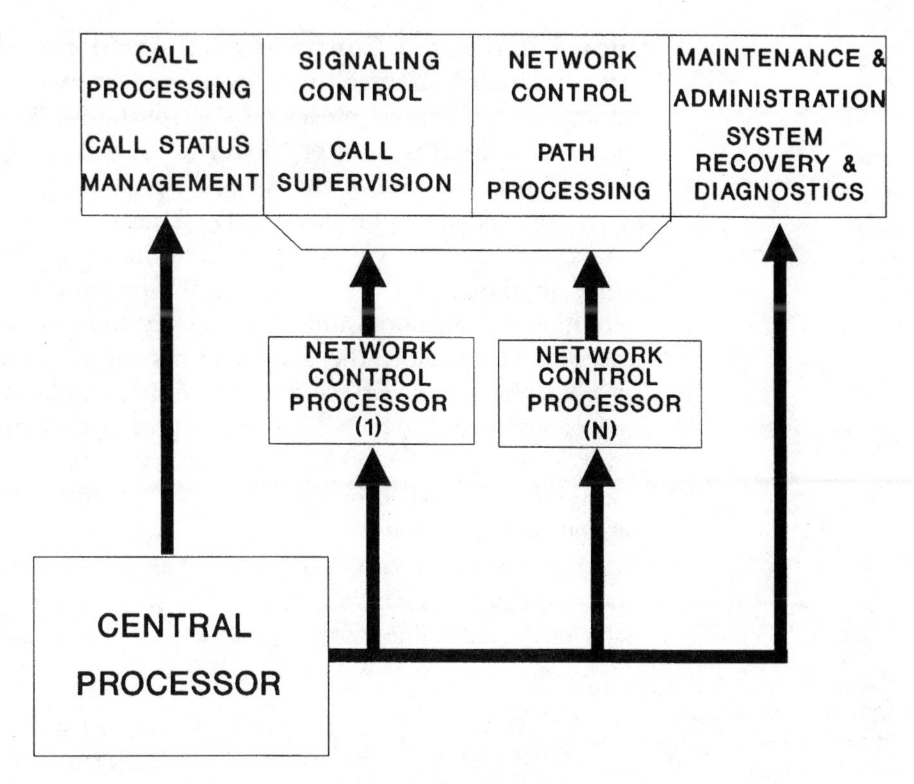

larity during growth, since call processing is still centralized. Under this architecture, the line and trunk modules and other peripheral subsystems operate under the NCPs. During call origination, line modules will scan lines under the control of NCPs. Once an off-hook condition is detected, the NCPs request validation information from the central processor. The central processor is required to keep its subscriber database updated and to conduct all translations of dialed digits. It is also required to keep a "map" of all calls being processed through the digital switching system.

In the case of system failure, the central processor is fully responsible for system recovery. All routing maintenance and diagnostics for the digital switching system are controlled by the central processor. All billing information is controlled by the central processor as well. Other responsibilities of the central processor include NCP diagnostics and, in the case of NCP failure, isolation of defective NCPs and control of the repair process. Depending on design specifications, the central processor may also assist in the local recovery process of the NCPs without affecting the operation of the entire digital switching system.

Quasi-Distributed Control-Based Architecture Currently the quasi-distributed architecture is being widely used (see Fig. 2.5). In this architecture, the central processors control the NCPs, and the NCPs control most of the functions of the digital switching system. In a variation of this architecture, one of the NCPs, designated NCP 0 in Fig. 2.5, may be given some added responsibilities. These responsibilities may include system recovery of other NCPs and maintaining the global image of all calls. In the quasi-distributed architecture, the line and trunk modules along with other peripheral modules are assigned to particular network control processors. For instance, the first 2048 lines and 64 trunks may be assigned to NCP 1, the next group of 2048 lines and 64 trunks to NCP 2, etc. In this type of architecture, each NCP has a preassigned periphery. The NCP maintains most of the information that is specific to the lines and trunks it serves. During call processing, the NCP validates subscriber lines and controls service circuits for its lines and trunks. However, the analysis of dialed numbers, routing, etc., is still the responsibility of the central processor. The NCPs have access to the switching fabric and can assign paths for the calls. The switching fabric may also be partitioned according

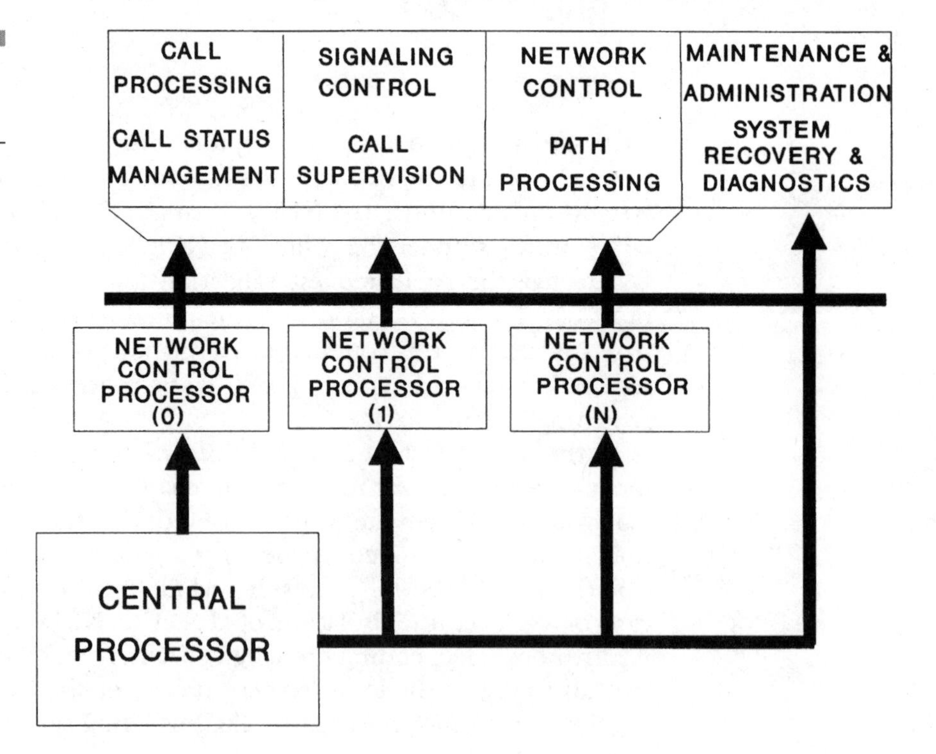

Figure 2.5 Quasi-distributed control

to particular switching architecture requirements. Details on partitioning are given later in this chapter.

In this architecture, some of the responsibilities of the central processor may be assigned to a particular NCP, say, NCP 0. This NCP communicates with other NCPs and may keep a global image of all calls through the system. It may also be made a guardian of other NCPs and may run routine diagnostics and aid in the recovery process of other NCPs. These functions are not shown in Fig. 2.5; however, this does not imply that other NCPs will not be required to maintain the call status information for their portions of the switching network. This strategy is clarified in later chapters, in which networking, partitioning, and control are covered in greater detail. Under this architecture, the central processor still has the overall control of all NCPs. During system recovery, the central processor will be responsible for overall system recovery process and will also be responsible for billing, system maintenance, and administration. The NCPs are required to implement switching functions autonomously for their portion of the switching network and are controlled by the central processor only when systemwide actions are required.

Distributed Control-Based Architecture A conceptual diagram of a fully distributed control architecture is shown in Fig. 2.6. This architecture can also be defined as a "central-processor-less" architecture. This architecture supports no central processor. All functions of the digital switching system are divided into smaller processing functions. These functions support all actions needed of a digital switching system but are contained in independent processing units. These processing units communicate with other units through messaging. When a subscriber makes a call, the associated line module processing unit gets involved with the call. In a distributed architecture, a line unit does much more than identify an off-hook condition of a subscriber. The line unit continuously scans the lines of its subscribers, keeps track of all its calls, maintains the database for its translation, conducts call supervision, and even conducts local recovery and local diagnostics of its circuitry. All processors communicate to and through the switching fabric via control messages. In theory, this is the ultimate in switching system design. But without the proper design, messages passing between various processors may cause bottlenecks and implementation of complex telephony features is difficult. However, these limitations are now being eliminated with the advent of faster processors, improved messaging protocols, and enhanced architectural techniques.

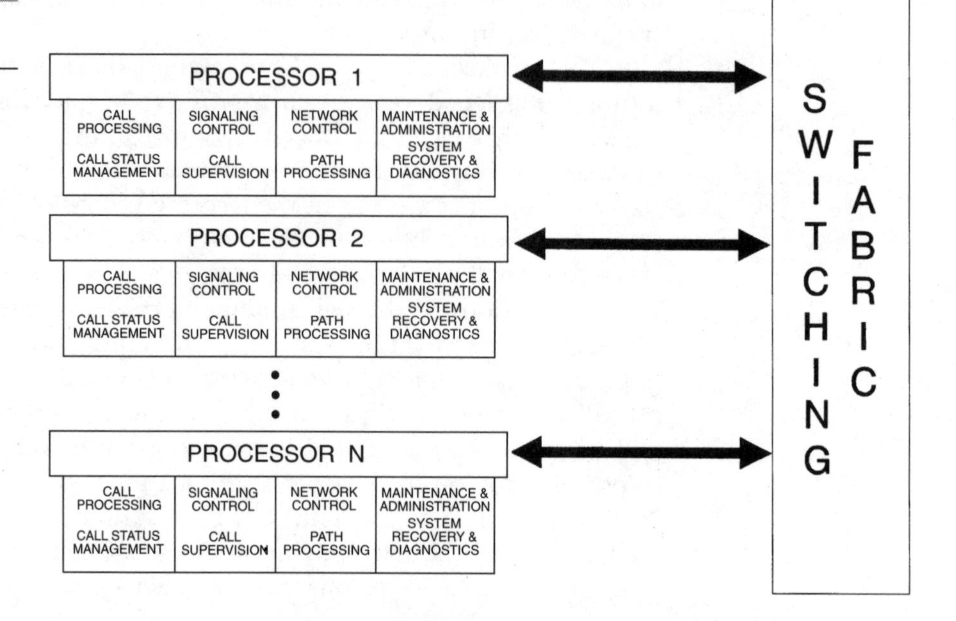

Figure 2.6 Distributed control

2.2.9 Multiplexed Highways

As mentioned before, this book assumes that the reader is familiar with various forms of digitizing techniques used in a digital switching system, such as pulse code modulation (PCM) and pulse amplitude modulation (PAM), or multiplexing techniques, such as frequency-division multiplexing (FDM) and time-division multiplexing (TDM). For more information on these and other techniques, consult the References at the end of this chapter. To understand the communication links between various parts of a digital switching system, the concept of a multiplexed highway is important. In short, a multiplexed highway is a digital link between different parts of a digital switching system carrying information in a time-multiplexed mode. Thus the incoming digitized information is encoded into time slots, and the outgoing data are decoded or extracted from the time slots. This multiplexing and demultiplexing is shown in Fig. 2.7*b*. One of the objectives of line and trunk modules is to provide concentration of lines and trunks entering the switch; this reduces the cost and size of a digital switching system. Different levels of concentration can be provided by the multiplexing and demultiplexing techniques. Different digital switching system subsystems may use different forms of digitizing and concentration techniques. But most of the multiplexing and demultiplexing is done in the peripheral elements of a digital switching system. The basic idea of multiplexed highways is shown in Fig. 2.7. The main objective of a digital switching system is to connect line, trunk, and peripheral modules through the switching fabric for the purpose of setting up a path for a line or a trunk connection. Since incoming and outgoing analog lines and trunks are still being used, analog-to-digital (A/D) converters are needed before the multiplexers and the digital-to-analog (D/A) converters after the multiplexers. Multiplexing and demultiplexing techniques are used in a digital switching system primarily to interface high-speed elements and low-speed elements and also for the cost reduction associated with digital switching systems transmission requirements.

The multiplexed highway of a typical digital switching system can generally be classified into three categories, shown in Fig. 2.7*c:*

- Subhighway
- Highway
- Junctor highway

Subhighway The link between the line and trunk modules and the multiplexer or demultiplexer is usually classified as a subhighway. Usually PAM, PCM, and similar type information are exchanged at the subhighway level.

Figure 2.7 Multiplexed highways

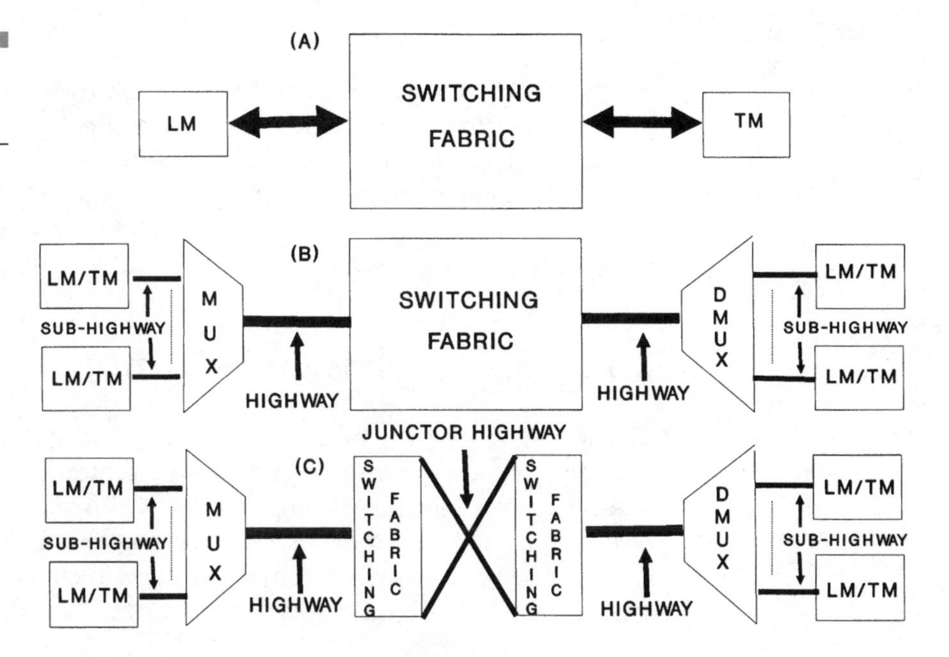

Highway The link from the multiplexer or demultiplexer to the switching network is usually called a *highway*. This highway is the main artery between the switching network and the multiplexers and demultiplexers. The information carried through the highway is in TDM format and generally is carried at high speed.

Junctor Highway At this stage, from a conceptual point of view, we assume that the junctor highway connects various portions of a partitioned switching network. The type of information carried through the junctor highway is similar to that on the highway.

2.3 Switching Fabric

The concepts of line-side and trunk-side switching functions were introduced in Chap. 1. This section expands the switching fabric portion of the digital switching system block diagram. The usual connections that a digital switching system is required to establish are

1. Line-to-line (L-L) connections
2. Line-to-trunk (L-T) connections
3. Trunk-to-line (T-L) connections
4. Trunk-to-trunk (T-T) connections

All these connections are established through the switching matrix of a digital switching system. Since this represents the basic "fabric" of a switch, the term *switching fabric* is sometimes used to describe the elements that establish network paths through a switch. In this section we introduce various forms of switching fabrics and switching schemes commonly employed in most modern digital switching systems.

2.3.1 Space-Division Switching

The space-division switching fabric is the oldest, and it was used by the electromechanical step-by-step and crossbar central offices long before the development of digital switching systems. It also was the switching fabric of choice for the early designers of first-generation electronic switching systems. The basic concept of this switching element is shown in Fig. 2.8. It consists of physical cross-points which can be connected via control sig-

Figure 2.8
Space-division fabric S

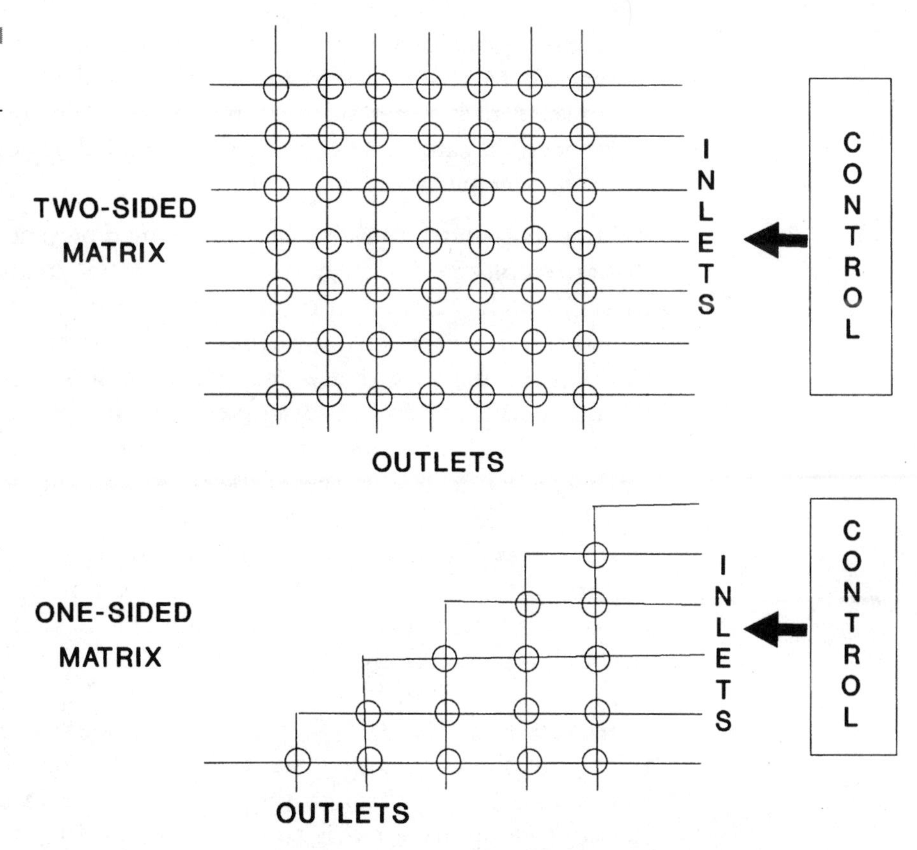

nals. For present-day digital switching systems, these control signals are provided by microprocessor-actuated controllers. In electromechanical systems, this type of control was provided by relay-based controllers. Currently, 1ESS*, 1AESS, 2ESS, and 3ESS switches employ space-division switching.

The S Switch Two basic configurations for the S switch are shown in Fig. 2.8:

- *Two-sided matrix.* This allows a two-sided connection of an outlet to an inlet. For instance, the connection of outlet 1 to inlet 3 and the connection of 1 to outlet 3 can be established simultaneously, thus allowing reciprocal connections.
- *One-sided matrix.* This allows only a one-sided connection between an outlet and an inlet; sometimes it is referred to as a *folded matrix.*

* ESS is a trademark of Lucent Technologies.

> In this type of arrangement, the redundant side of the fabric is removed. More sophisticated types of controllers are needed that can keep track of one-sided connections since reciprocal connections cannot be supported by a folded or nonredundant matrix scheme.

These two configurations by no means represent the only schemes that are employed currently. A variety of other arrangements are often used in which the amount of network availability can be predicted by eliminating certain cross-points.[2] The S switch connects a path through the network that is maintained throughout the duration of the call, and the T switch (discussed later) maintains a path only during a specified time slot. The S switch also provides a "metallic" or a real metal connection between the inlets and the outlets, whereas a T switch provides a path through the network via memory assignments. Different types of connections through the switching network require different numbers of cross-points. For instance, a four-wire trunk connection that needs two simultaneous paths will require two cross-point connections.

The impact of an S switch on a time-multiplexed bit stream, also referred to as a *time slot* (*TS*), is shown in Fig. 2.9. The left-hand side of the figure shows various time-slot intervals labeled TS 1, TS 2, TS 3. The trace of a path through the S switch of the TS 1 time slot follows; the contact points for this time slot are shown by solid circles. The control signal actuates contacts 1, 2 and *N*, 5. As a result of this operation, the TS 1 signal entering inlet 1 will depart outlet 2, and similarly the signal entering inlet *N* will depart outlet 5 for this time interval. This example illustrates that the S switch provides a metallic path through the switching fabric for a known interval of time.

2.3.2 Time-Division Switching

The time-division switching fabric is now a de facto standard for designing modern digital switches. The most important advantage of the time switching fabric, besides lower cost, is that unlike space-division switching fabric, it allows sharing of the cross-points. A conceptual illustration of a typical time-division fabric is shown in Fig. 2.10. The concept of time division has been around for years, mostly employed in transmission products. Its use as a switching fabric is more recent. The time-division switching fabric can be considered to

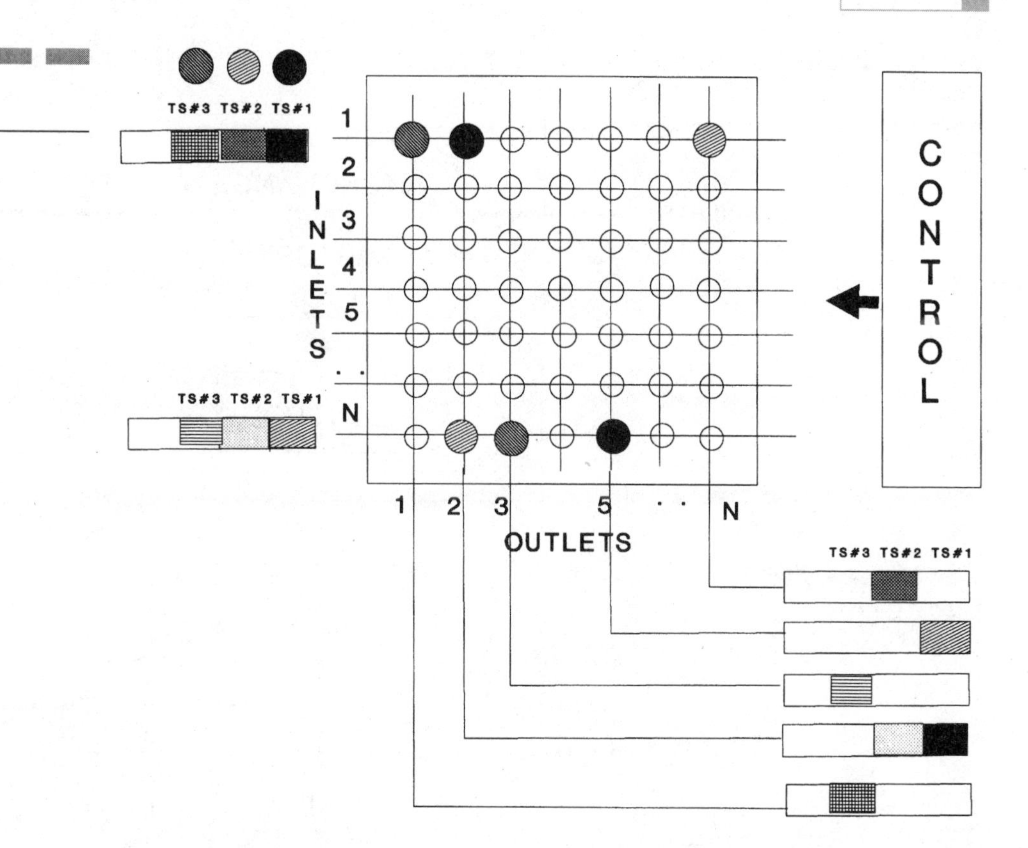

Figure 2.9
S switch

be a memory system that assigns different memory locations for different time slots, and it is referred to as *time slot interchange* (*TSI*) memory. This type of "soft" assignment allows sharing of cross-points for short periods. The memory allocation or TSI is controlled by fabric controllers.

The T Switch The T switch is currently considered to be the basic and important element of the digital switching system's switching fabric. The basic concept of the T switch is shown in Fig. 2.11. There are N inlets and N outlets, numbered 1 to N. Assume that time slots TS 2 and TS 1 are entering inlet 1; time slots TS 22 and TS 21 are entering inlet 2; etc. Since the TSI scheme is nothing more than a memory rearrangement system, complete flexibility in reassignments of different time slots to different outlets can be accomplished via controller commands, as shown. This type of TSI reassignment is done continuously during the duration of a call, in effect, allowing sharing of cross-points and hence making the switching fabric more economical.

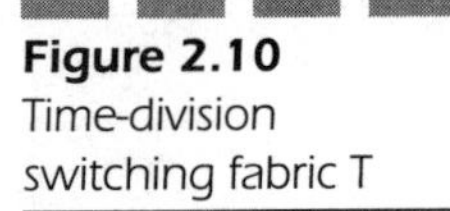

Figure 2.10
Time-division switching fabric T

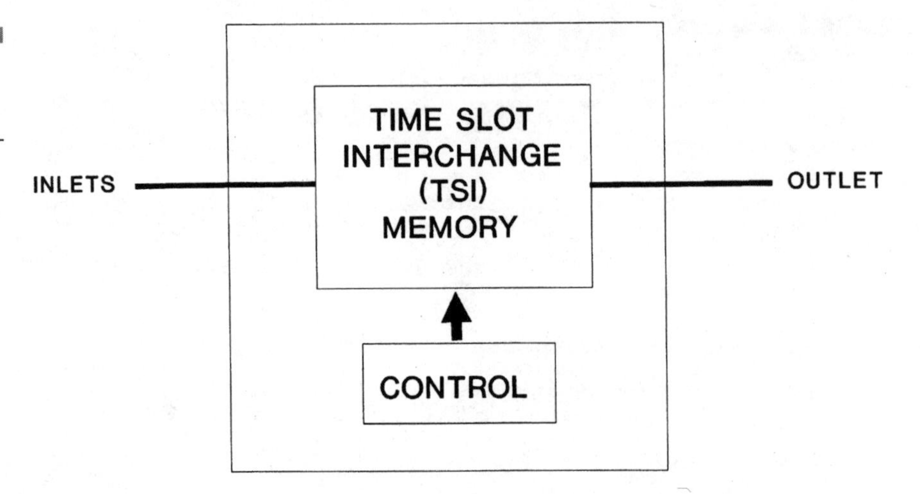

Figure 2.11
T switch

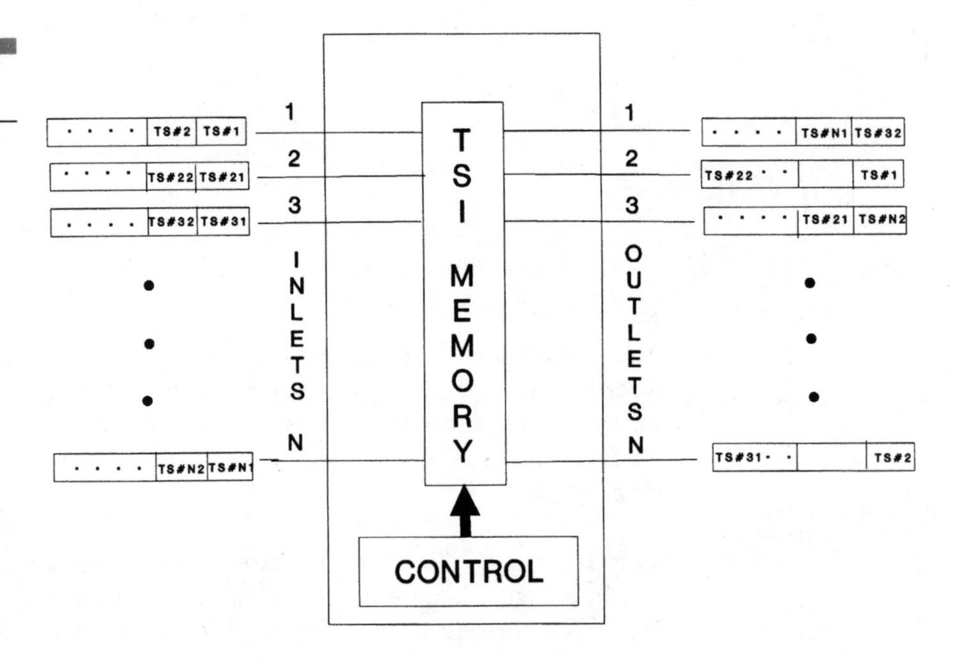

2.3.3 Space-Time-Space (STS) Switching

One objective in the design of a modern digital switching system is to reduce costs and improve the switching efficiency of the fabric. Obviously, there is a practical limit to the size of a single switching stage that can be effectively utilized. At present, various combinations of S switches and T switches are used to accomplish the above objective. One combination uses an S switch followed by a T switch and a final S switch. This arrangement, referred to as *STS fabric*, is shown in Fig. 2.12. This particular arrangement depicts $N \times M$ (meaning N inputs and M outputs) size, with N S switches separated by M T switches. In an STS switching fabric, a path through the network is established via smart network controllers that link an incoming time slot with an outgoing time slot. This type of time slot linkage is then dynamically updated throughout the duration of a call.

2.3.4 Time-Space-Time (TST) Switching

One of the most popular switching fabric arrangements currently deployed by digital switching systems is based on time-space-time (TST) architecture, as shown in Fig. 2.13. An incoming time slot enters a T switch; a path is hunted through the S switch for an appropriate outgoing time slot; and once identified, the path through the switching

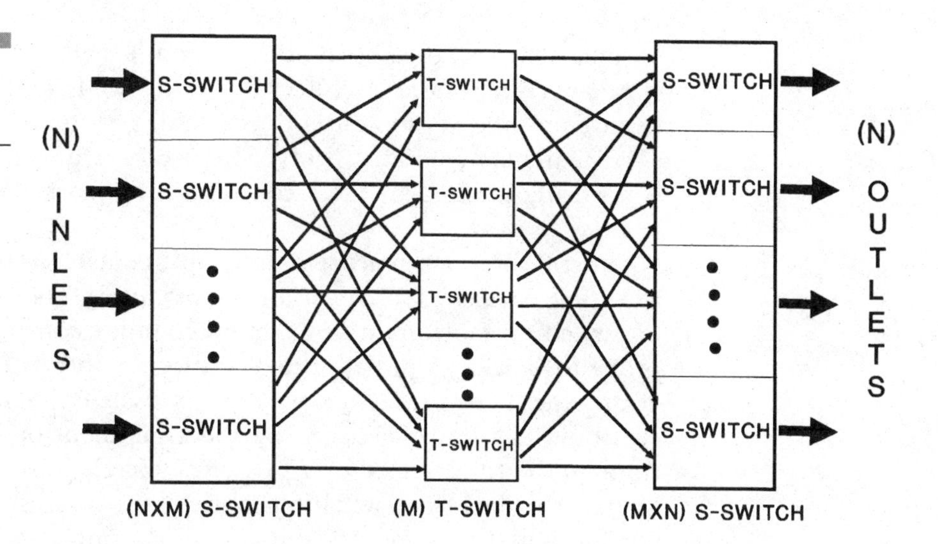

Figure 2.12 Space-time-space (STS) switching

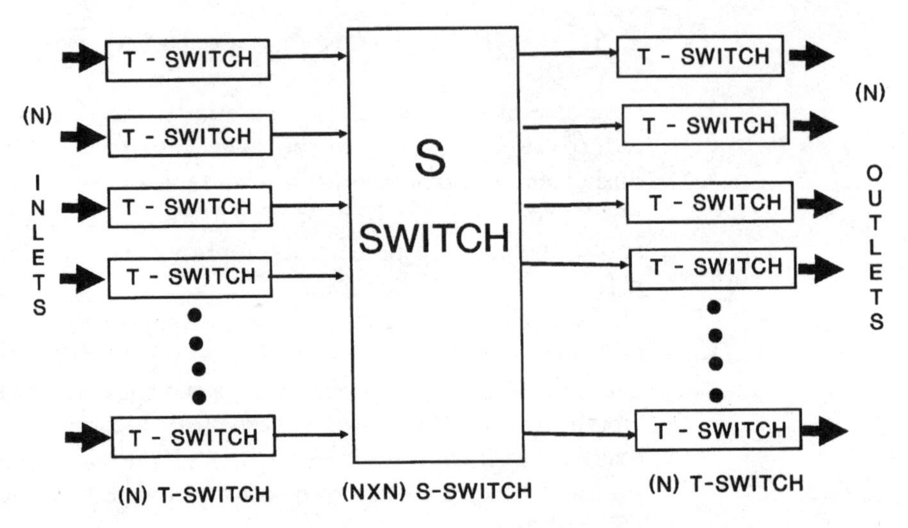

Figure 2.13 Time-space-time (TST) switching

fabric is established and dynamically updated throughout the duration of the call. One of the basic advantages of the TST architecture over the STS architecture is that it can be implemented at a lower cost, since T switches are less expensive than S switches and under heavy traffic offer more efficient utilization of time slots with lower blocking probabilities.

2.3.5 Time-Time-Time (TTT) Switching

Various other arrangements using S and T switches are possible in a large switching fabric. Architectures using TSST, TSTST, TTT, etc., have been employed in some digital switching systems, and many other arrangements are under study. The use of a particular architecture is normally dictated by costs, controller complexity, and traffic characteristics.

One of the least expensive but implementable arrangements is the time-time-time (TTT) architecture, shown in Fig. 2.14. However, this type of switching fabric requires a more complex controller and may be prone to blockages during heavy traffic. As shown in the figure, the arrangement requires three stages of T switches. A path through the switching fabric is established by the assignment of proper time slots through the three stages for a particular speech path that allows speech to pass through the switching fabric. This path is then maintained dynamically throughout the duration of the call.

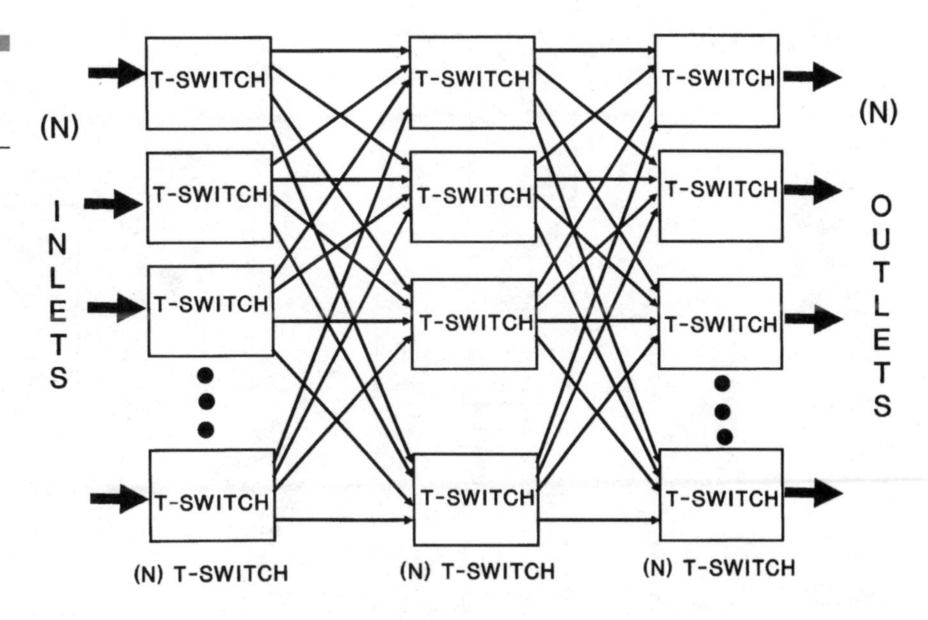

Figure 2.14 Time-time-time (TTT) switching

2.4 Programmable Junctors

Sometimes very large switching network fabrics become necessary, or an existing network requires growth. Usually, a junctor arrangement or simply junctors are employed for this purpose. A junctor is a link between network fabrics. An example of a junctor arrangement is shown in Fig. 2.15. The junctor connections can be manually or automatically programmed to connect one network to another network depending on the connection requirements. If the number of inputs is larger than the number of outputs, then the arrangement is classified as a *concentrator*; but if the number of inputs is smaller than the number of outputs, then the arrangement is classified as an *expander.* Programmable junctors are primarily used during the growth of an existing digital switching system. Junctors enable the growth process without replacing existing networks.

2.5 Network Redundancy

As mentioned earlier, the architecture of a modern digital switching system calls for redundancy for most of its subsystems. One of the subsystems not duplicated is the line and trunk modules, but the switching

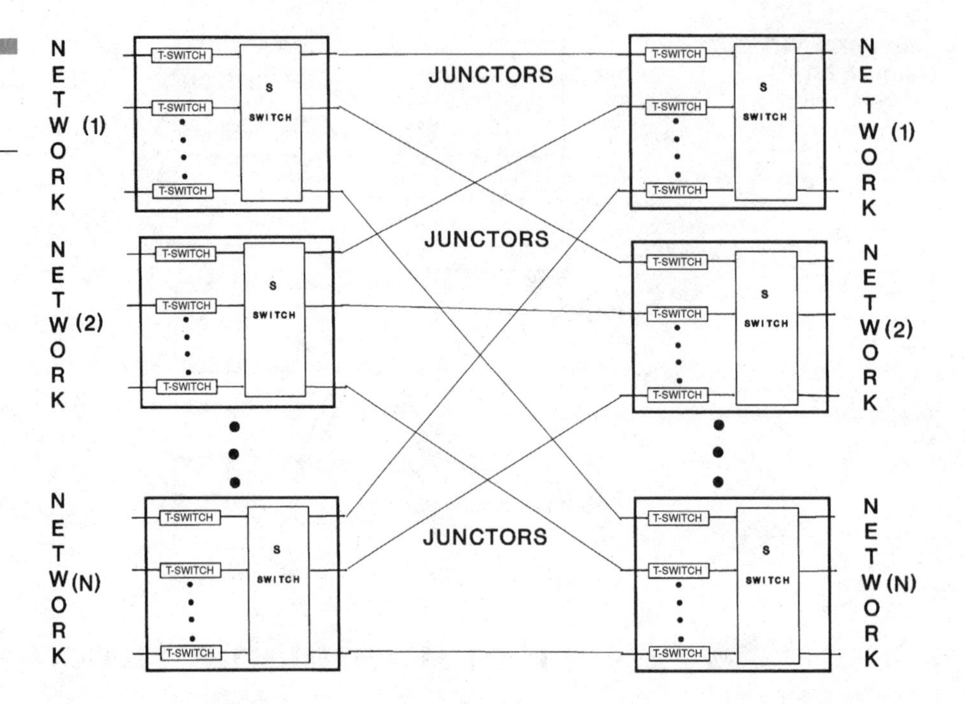

Figure 2.15 Programmable junctors

network to which they connect sometimes are duplicated; refer to Fig. 2.16 for such a scheme. Various types of redundancy schemes are currently employed. The basic objective is to provide redundant fabric connection for an established call through the network. In the event of a multiplexer, highway, or other type of failure that impacts the connection path of established calls, an entire redundant network could be switched in to improve the reliability of established calls.

2.6 Summary

This chapter introduced the basics of communications and control associated with current digital switching systems. Various levels of controls were explained along with basic central functions. Different classifications of digital switching system control architectures and the concept of multiplexed highways were also discussed. This chapter also presented the basic concepts related to two basic types of switching fabric, the S switch and T switch, and various combinations of these switches were developed. The concepts of junctors and network redundancy were introduced.

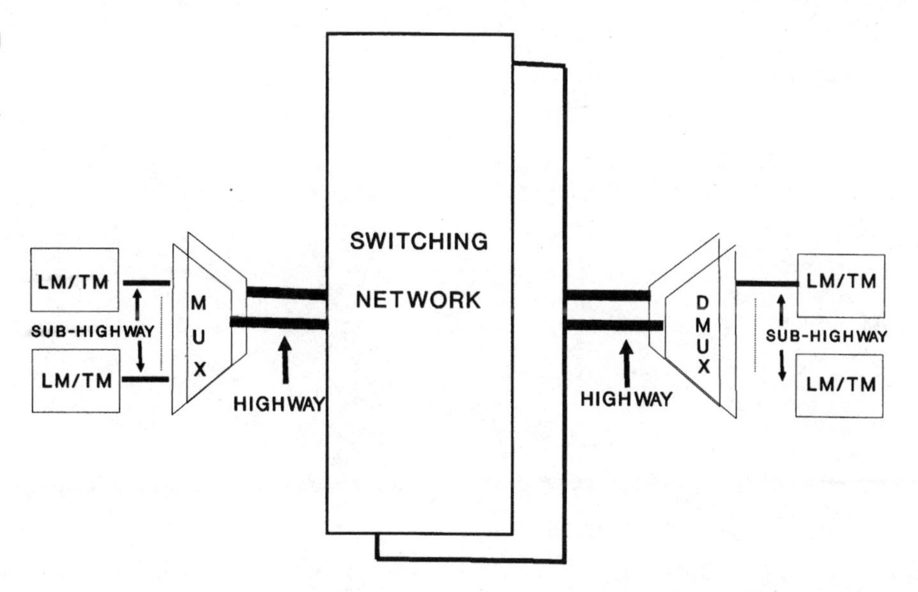

Figure 2.16
Network redundancy

REFERENCES

1. Syed R. Ali, "Implementation of Firmware on SPC Switching Systems," *IEEE Journal on Selected Areas in Communications,* October 1988.
2. John Bellamy, *Digital Telephony,* Wiley, New York, 1982, Chap. 5.

CHAPTER 3

Reliability Modeling

3.1 Introduction

The first two chapters introduced the reader to the building blocks and control structures of digital switching systems. The approach taken was an inside-out one, and the advantage of this approach will now become apparent. A digital switching system is very complex and large, and so assessing its reliability can be complex and difficult. However, a reasonable assessment of system reliability can be made if one approaches the problem with good and proven analytical techniques. This chapter covers methodologies that could aid in assessing hardware reliability. Software reliability is covered in later chapters. This chapter assumes that the reader is familiar with some basic concepts of hardware reliability; however, some essential basics are covered. If the reader feels a need for more details, an exhaustive reference list is provided at the end of this chapter. At this stage, the reader should be familiar with the basic subsystems of a modern digital switching system. We now introduce some essential reliability concepts with enhancements that include redundant schema of a digital switching system.

3.2 Scope

This chapter covers some basic concepts of reliability with emphasis on Markov modeling via state transition diagrams. Associated concepts required for analyzing the reliability and downtimes of digital switching systems are also covered. We describe concepts related to failure models, failure rates, and repair times that are appropriate for digital switching system analysis. The basics of sensitivity analysis are also discussed.

3.3 Downtimes in Digital Switching Systems

The main purpose of a reliable digital switching system is to provide telephone service without interruption or degradation for long periods. As a matter of fact, the telecommunications industry standards[1] and practice[2] require that a telephony switching system not be nonoperational, referred to as downtime, for more than 3 min/yr or 2 h in 40 years! To design and maintain such a highly reliable system requires special attention to systems engineering and reliability.

As mentioned earlier, this book uses an inside-out approach. The following sections introduce various aspects of system reliability techniques that can be employed internally to predict and enhance the reliability of digital switching systems.

3.4 Purpose of Reliability Analysis

The basic objectives of reliability analysis are

1. To ascertain the reliability figures for a digital switching system and compare them with established objectives
2. To study the switching system architecture from a reliability point of view and suggest improvements if the present architecture is found to be less reliable than anticipated
3. To better understand the interaction of hardware and software functions that are related to system recovery, diagnostic, and repair methodologies

3.5 System Reliability Assessment Techniques

There are many system reliability assessment techniques. Some well-known techniques are

- Failure tree analysis
- Reliability block diagram analysis
- Markov-chain-based analysis

3.5.1 Failure Tree Analysis

This is one of the earliest methodologies used for determining the root cause of a catastrophic failure. A tree structure of a system is generated with different modes of failure assigned probabilities for each type of failure. Chances of failure for the conceived scenario can then be calculated. Failure tree analysis is a viable technique that is still being used by the Department of Defense (DOD) and the National Aeronautics and Space Administration (NASA). This technique has not been used extensively for the analysis of digital switching systems.

3.5.2 Reliability-Diagram-Based Analysis

In this technique, the reliability block diagrams are constructed for each component, and depending upon their connected configuration, the overall system reliability is ascertained. For example, if components are connected in series and their reliabilities are R_1, R_2, R_3, ..., R_n, then the overall reliability of the system is

$$R_{sys} = R_1 R_2 R_3 \cdots R_n$$

Thus the product rule applies. If the components are connected in parallel, then

$$R_{sys} = 1 - (1 - R_1)(1 - R_2)(1 - R_3) \cdots (1 - R_n)$$

where $1 - R_1$, $1 - R_2$, ... represents the unavailability of the subsystems, which can be approximated by

$$R_{sys} = R_1 + R_2 + R_3 + \cdots + R_n$$

Thus the addition rule applies.

In other words, if one of the units fails, the other units can still function and their failure rates are assumed to be independent of one another. A complex system could be decomposed into groups of parallel and serially connected blocks, and the system reliability can be assessed based on the interconnection of these groups. This methodology is useful in assessing the reliability of smaller subsystems that do not have very complex subsystem dependencies or interconnections. For the purpose of digital switching system analysis, the easiest method of determining the digital switching system reliability is based on Markov chain analysis.[3] This method has limitations, since it assumes fixed (exponential) repair and failure rates, but it is a powerful tool for calculating the downtimes of digital switching systems. To understand this technique, one has to understand the concept of state transition diagrams. In cases where repair rates are nonexponential, i.e., they are time-dependent, approximate Markov models can be used. This type of model is more complex but can provide greater accuracy.[4]

3.5.3 System Reliability Diagram

At this stage let us construct a system diagram for the digital switching system developed so far. Figure 3.1 shows a very simple system reliability diagram. This diagram is based on redundancy for the interface controllers (ICs), network control processors (NCPs), and the central processor (CP) and no redundancy for the line or trunk modules. Notice this diagram is modeling the processor units in duplex modes and the line and trunk modules in a simplex mode. Studying this diagram reveals that the line and trunk modules do not have any redundancy and are connected in series to the ICs. The ICs are duplicated, and therefore two ICs are shown in parallel. This redundant IC subsystem is shown connected to the line and trunk modules on one side and to the NCPs on the other side. There may be a number of NCPs, and multiple NCPs may be shown in such a diagram. But for this simple example, consider just one duplicated NCP pair. This duplicated NCP pair connects to the CP. The CPs are also duplicated. This represents the simplest reliability diagram for the main components of the digital switching system. This diagram is enhanced in the next chapter.

3.5.4 Markov-Chain-Based Analysis

The Markov chain was established by the Russian mathematician Markov early in this century. He developed this probability model

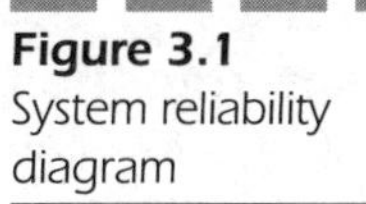

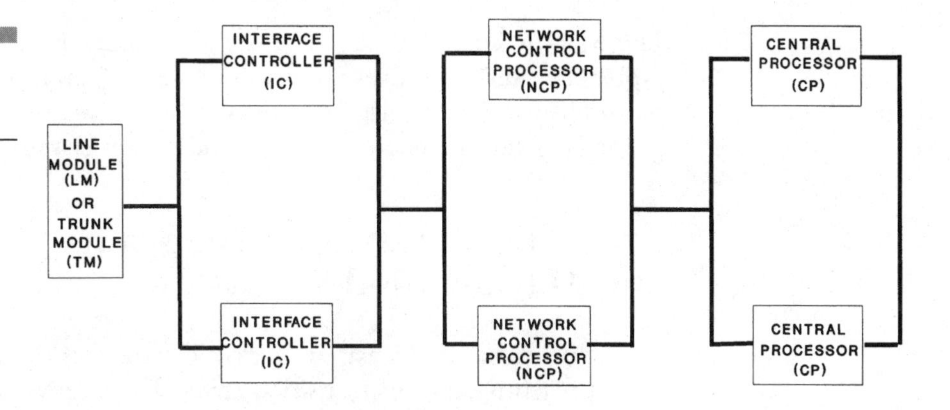

Figure 3.1 System reliability diagram

based on successive outcomes of trials that are dependent on each other and on immediate predecessor trials. The basic Markov process depends on the state of a system and its transitions. Markov chain analysis has been successfully applied in many diverse fields including physics, economics, operations research, and control engineering, to name a few. When this process is extended to a new field of application, it is important to clearly identify the states and the transitions of states. For digital switching systems, each operational and failure state needs to be identified with corresponding transitions defined.

A discrete-time Markov chain is based on a stochastic (time-varying probabilistic) process with independent variables. A Markov chain represents a sequence in which dependent events need to go back only 1 unit in time. This implies that the future condition of a process is dependent on only the present state, and not in its history. This is referred to as a *markovian property.*

The stochastic process $\{X_n, n = 0, 1, \ldots\}$ with state space I is called a discrete-time Markov chain[5] if, for each $n = 0,1,\ldots,$

$$P\{X_{n+1} = i_{n+1} | X_0 = i_0, \ldots, X_n = i_n\} = P\{X_{n+1} = i_{n+1} | X_n = i_n\}$$

$$\text{for all possible values of } i_0, \ldots, i_{n+1} \in I.$$

For a Markov chain with time-homogeneous transition probabilities, it is

$$P\{X_{n+1} = j | X_n = i\} = p_{ij} \qquad i,j \in I$$

independent of the time parameter n. The probabilities p_{ij} are called *one-step transition probabilities* and satisfy

$$p_{ij} \geq 0, \qquad i, j \in I \qquad \text{and} \qquad \sum_{j \in I} p_{ij} = 1, \qquad i \in I$$

The Markov chain $\{X_n, n = 0, 1, \ldots\}$ is completely determined by the probability distribution of the initial state X_0 and one-step transition probabilities p_{ij}. For a rigorous mathematical treatment of the Markov process, consult the references at the end of this chapter.

3.5.5 State Transition Diagrams and Markov Models

The solution of the transition matrix is based on the probabilities of a system reaching various transition states. The steady-state Markov chain can be solved by many different techniques, including flow rate, direct methods [for example, the Grassmann, Taksar, and Heyman[6] (GTH) technique] and iterative methods.[7] The solution to the steady-state distribution of a Markov chain is shown in Fig. 3.2*a* and is

$$\sum_{n \neq m} P_n \lambda_{nm} = P_m \sum_{n \neq m} \lambda_{mn} \qquad \text{under constraint} \sum P_n = 1$$

Conceptually, Fig. 3.2*a* shows a state *m* being entered with *N* inputs having various probabilities from P_1 to P_n and failure rates between λ_{1m} and λ_{nm}. The approximate outputs predicted by the flow rate equation are $P_1\lambda_{1m}$ to $P_n\lambda_{nm}$, under the constraint that the sum of all the probabilities must be unity.

Figure 3.2*b* shows an example of two-outcome model for splitting probabilities associated with a single state. In essence, probabilities assigned to a state can be split between various outputs and inputs if the sum of all the probabilities is kept to unity.

Failure Rate A failure rate is the rate at which hardware or a software component fails and is usually designated λ; it may have subscripts that

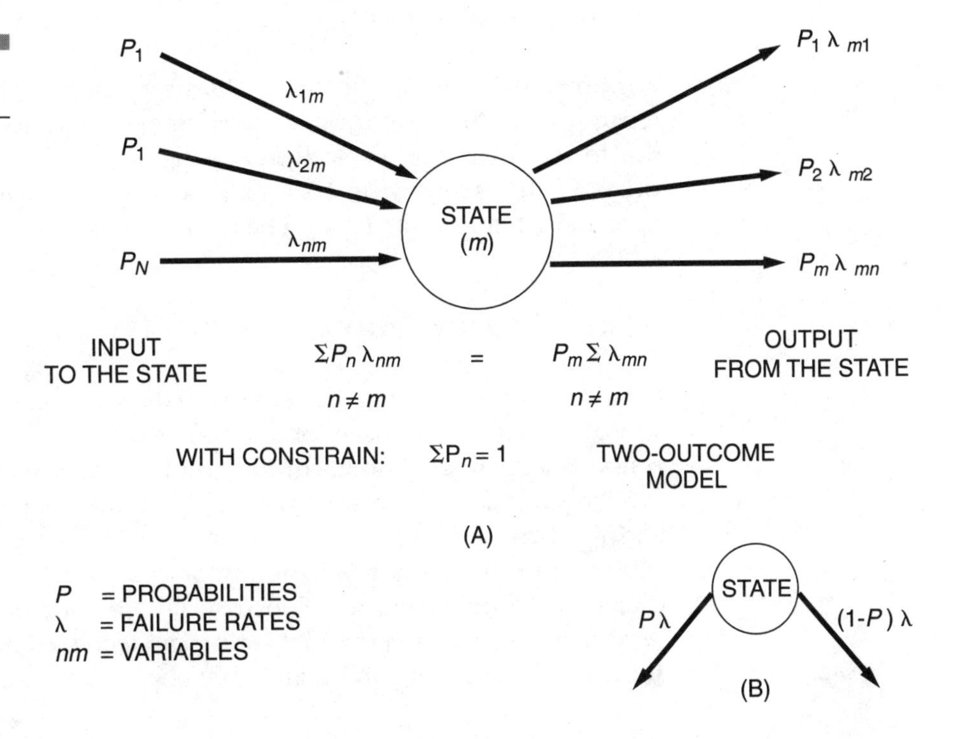

Figure 3.2 Flow rate model

designate various subsystems or states. Many of the reliability models use the failure in 10^9 h (FIT) rate.

The determination or prediction of the failure rate accurately plays a very important role in predicting the reliability of a digital switching system. This in itself is a science, and to cover this topic in detail goes beyond the scope of this book. Usually the U.S. telecommunications industry uses Bellcore's reliability prediction procedure (RPP).[8] The RPP is based on three methods for predicting circuit pack failure rates:

- Method 1 is based on the parts count. It adds the failure rate of each component on the circuit pack.
- Method 2 is based on statistical estimates and uses date from laboratory tests.
- Method 3 is based on statistical estimates of in-service reliability and field tracking studies.

After the failure rate of all circuit packs is determined, the failure rate of a hardware module is established by adding the failure rates of all circuit packs in that module. Similarly, the failure rate of a subsystem which may comprise a number of modules is determined by adding all failure rates of the modules. This information is then used for reliability analysis at the subsystem level.

Repair Rate The repair rate is the mean time in hours required to repair a fault; this may include traveling time required by a repair person to reach the site. Typically, for a digital switching system, a repair time of 3 h is required to fix a hard fault, which includes 2 h of travel time and 1 h for on-site repair. The repair rate is usually designated μ.

3.5.6 Simple Markov Models

Consider a state transition diagram that is drawn to represent various operational/nonoperational states of a system, shown in Fig. 3.3. This example considers a duplicated subsystem community connected in parallel.

Such subsystems are commonly employed in digital switching systems and represent a duplicated or redundant architecture. For instance, this subsystem could easily represent any of the duplicated subsystems of a digital switching system developed so far. For our present purpose of developing basic concepts of reliability modeling, consider that the two subsystems operate in parallel and only one is active or is in control of its

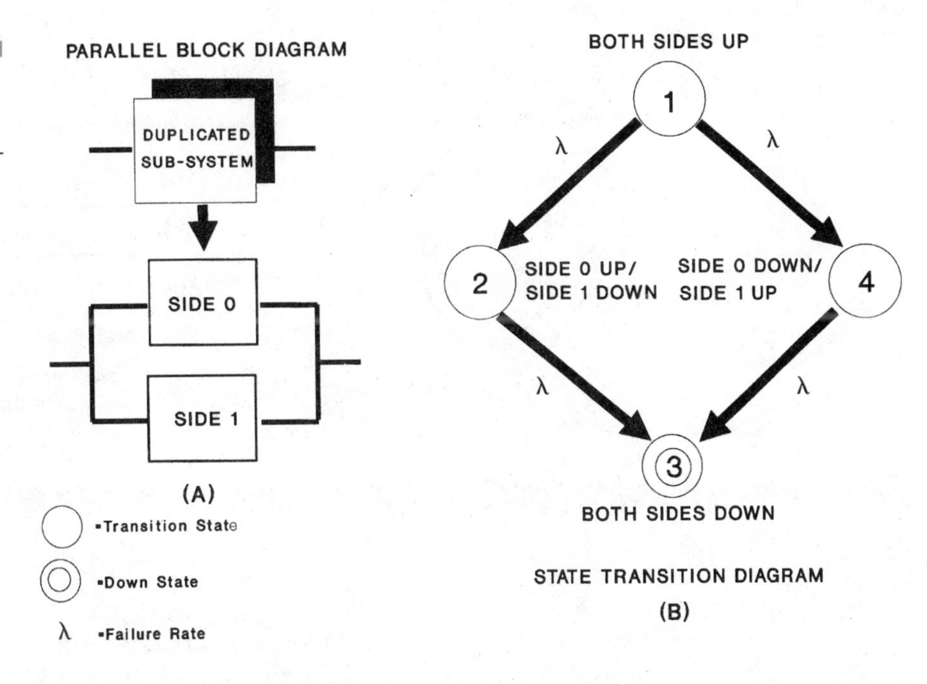

Figure 3.3 Markov model for a duplicated subsystem

functionality. Let us designate one subsystem as side 0 and the other as side 1. If both sides are operational or "up," we call it as the normal or *duplex* state and designate it as transition state 1. The designation of state numbers is arbitrary, but the type of state defined is important and should be noted on the diagram. Consequently, if either processor is not operational or is "down," then different states are designated to them; in this case states 2 and 4 represent the simplex states and λ represents failure rate. And if both the processors are down, then a *failed* state exists and in this example is designated as state 3 enclosed by double circles. This in essence is the Markov model based on transition states for the duplex subsystem community which is connected internally in parallel. In the next few sections we explain the Markov chain process in greater detail.

Markov Model of a Parallel Arrangement The transition diagram for a parallel-connected subsystem with a repair rate μ is shown in Fig. 3.4. Assume, the system was originally working in duplex mode, shown as state 1. One of the processors fails, and the system enters simplex-mode operation, shown by states 2 and 4. Figure 3.4*b* is simply a folded representation of Fig. 3.4*a*. And let us now assume that the processor working in simplex mode also fails; then the processor subsystem reaches a down state. This example assumes that the entire system will reach a down state

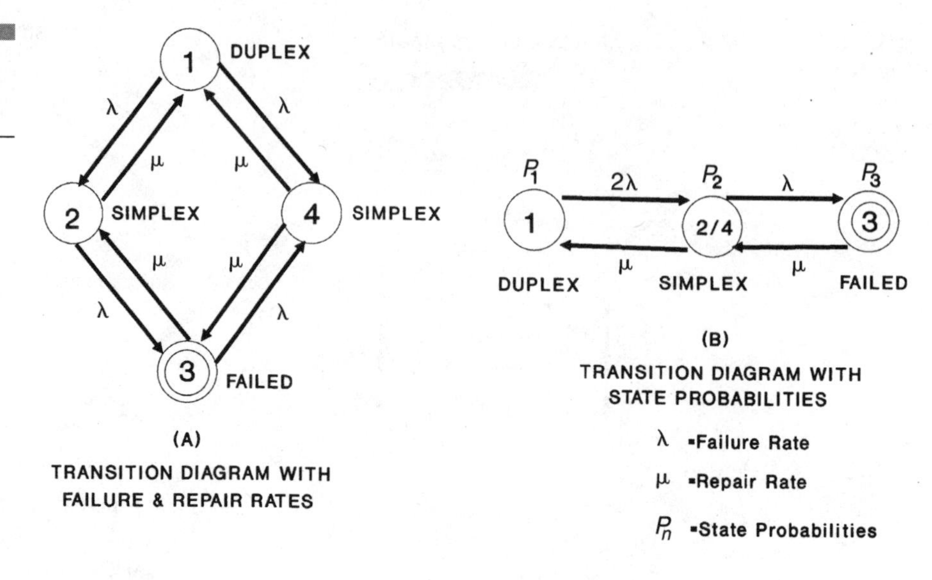

Figure 3.4 Parallel architecture modeling

if both subsystems go down, shown as state 3. Usually, the subsystem will be repaired, a processor at a time. In other words, the repair process will bring the system first to a simplex state and then to a duplex state, as shown in Fig. 3.4*b*.

Markov Model of a Series Arrangement As shown in Fig. 3.5*a*, side 0 of the subsystem for this example can be further decomposed into two elements connected in series. The Markov model for side 0 as well as for side 1, assuming symmetry, is shown in Fig. 3.5*b*, and a simplified Markov model is seen in Fig. 3.5*c*. Notice the difference between the two Markov models in Figs. 3.4*b* and 3.5*c*. They are essentially the same except in the number of down states. In a series arrangement, the entire side will fail in the case of failure of either element; in a parallel system, both sides have to fail to create a down state.

A Flow Rate Solution This subsection shows a manual solution of a Markov model using the flow rate equation. In practice, for models that contain more than a few states, manual solution of the Markov model is not practical, and computer-based tools should be employed. As shown in Fig. 3.4*b*, the probabilities assigned to each transition state are

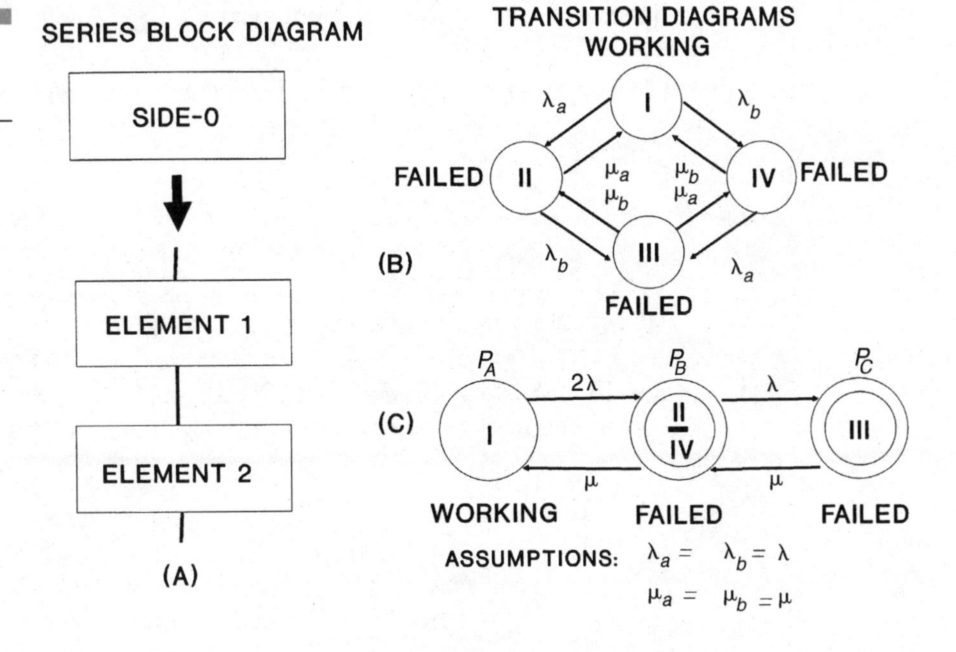

Figure 3.5 Series architecture modeling

Probability	State
P_1	1
P_2	2, 4
P_3	3

Now we apply the flow rate equation to the above model. Three equations are obtained for each state:

For state 1:

$$P_1 2\lambda = \mu P_2$$

For state 2:

$$P_2(\lambda + \mu) = 2\lambda P_1 + \mu P_3$$

For state 3:

$$P_3\mu = \lambda P_2$$

Solving the above equations, one obtains

$$P_1 = \frac{\mu^2}{\Theta} \qquad P_2 = \frac{2\mu\lambda}{\Theta} \qquad P_3 = \frac{2\lambda^2}{\Theta}$$

where

$$\Theta = 2\lambda^2 + \mu^2 + 2\mu\lambda$$

Use of this equation with real values and computerized solution of similar equations are discussed later in this chapter.

Solving Markov Chains In Markov chain analysis, one of the most common equations[5] that needs to be solved is a set of linear equations

$$x_i = \sum_{j=1}^{N} p_{ji} x_j \qquad i = 1, \ldots, N$$

where $\sum_{i=1}^{N} x_i = 1$ has a unique solution.

Two methods can be used to solve this set of equations: direct methods and iterative methods.

One common method of solving this type of equation is to use gaussian elimination, such as the Gauss-Jordan method. This method is viable as long as the number of equations does not exceed 200 or so. Software for solving this type of equation is commercially available. This method requires large storage space in computer memory, and the computational effort is proportional to the cube of the number of equations (N^3).

For really large systems, iterative methods with successive overrelaxation have been used to solve Markov chains. The Gauss-Seidel method is a special case of successive overrelaxation which can be used for this purpose.

However, the author found the use of the Grassmann, Taksar, and Heyman (GTH) technique (a direct method) to be fast and accurate enough to calculate transitional probabilities for models of digital switching systems. Following is a high-level description of this technique. This technique requires about $2/(3N^3)$ computational effort. Advantage can be taken of specific structures that large problems may exhibit. For example, the authors of the GTH algorithm modified it for a tridiagonal block structure of queuing models and solved problems with 3600 states in reasonable time. This is the algorithm:

Let P be the transition matrix of a Markov chain with states $0, 1, \ldots, N$, and let x be a stationary distribution of P. Then

$$x = xP \qquad \text{and} \sum_i x_i = 1 \qquad \text{for all values of } x$$

where x is the matrix and x_i is an element of matrix x. Then GTH algorithm states:

1. For $n = N, N-1, \ldots, 1$, do the following: Let

$$S = \sum_{j=0}^{n-1} p_{nj}$$

$$p_{in} = \frac{p_{in}}{S} \qquad i < n$$

$$p_{ij} = p_{ij} + p_{in} p_{nj} \qquad i, j < n$$

2. Let

$$\text{Total} = 1 \qquad \text{and} \qquad x_0 = 1$$

where Total is a variable that holds changing values of Total.

3. For $j = 1, 2, \ldots, N$, do the following: Let

$$x_j = p_{0j} + \sum_{k=1}^{j-1} x_k p_{kj}$$

$$\text{Total} = \text{Total} + x_j$$

4. Let

$$x_j = \frac{x_j}{\text{Total}} \qquad j = 0, 1, \ldots, N$$

The solution of transition matrix P is thus obtained.

GTH Algorithm Accuracy All the calculations in Chap. 4 will be based on computation by the GTH algorithm. The author found the algorithm to be resistant to rounding errors and accurate to at least seven decimal digits. Of course, any computational technique may be used to calculate the transition metrics, but the accuracy of these techniques needs to be validated before large-scale modeling is attempted.

3.6 Failure Models

There are some specific failure Markov models[9] that are applicable to digital switching system architectures. These models may also apply to other network elements like signal transfer points (STPs), service control points (SCPs), and so on, that have common characteristics as shown below.

- *Detection failure:* A digital switching system fails to detect failure when it is supposed to.
- *Coverage failure:* A digital switching system fails during a switchover between active and standby modes.
- *Diagnostic failure:* A digital switching system's diagnostic cannot correctly identify failed components.
- *Recovery failure:* A digital switching system's emergency recovery program cannot bring the system back to an operational mode.

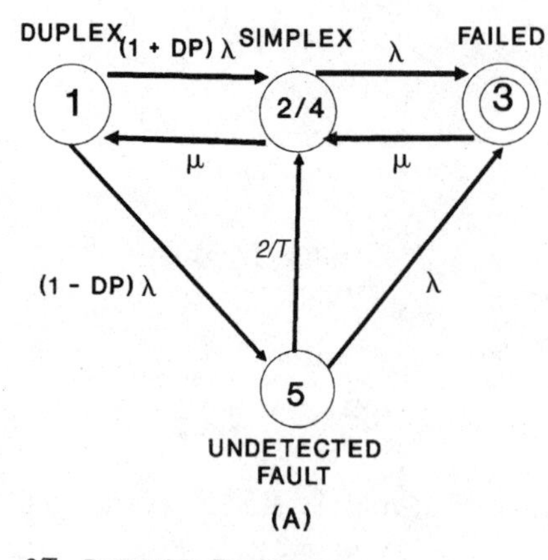

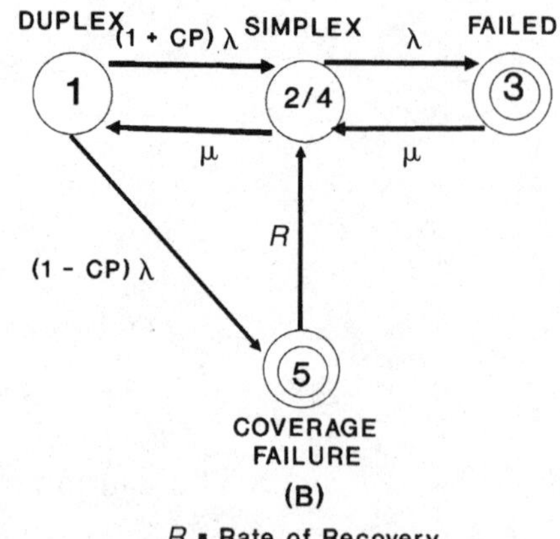

Figure 3.6 Detection and coverage failure models

3.6.1 Markov Model of a Detection Failure

This subsection introduces some simple concepts necessary for applying Markov chains to the analysis of digital switching systems. Clearly, a digital switching system performs a very complex real-time switching function and on many occasions may not perform exactly as expected. For instance, until now, all our models have assumed that in a duplex arrangement if one subsystem fails, then the standby or redundant subsystem takes over the function of the failed unit and instantaneously starts working correctly. However, in the real world, this does not always happen; hence the concept of detection failure must be understood and modeled. The detection failure probability is the failure of the system to detect faults in the standby system before switchover takes place and the standby system fails at switchover. A Markov model of a duplex subsystem showing a detection failure is seen in Fig. 3.6*a*. Note that this is only a model for the detection failure. A complex system may have a far more complex model for this type of failure or may even have multiple models for different conditions of failure. Figure 3.6*a* assumes that the frequency with which the system can detect the failure in the standby system is $2/T$. In other words, for a digital switching system a midnight routine or set of diagnostic programs is run once in 24 h, and $2/T$ represents this average frequency of detection. If something happens to this unit before the midnight routine and a problem goes undetected, the probability of this detection failure is labeled DP in the figure. The model also shows the addition of a new state to the duplex model as transition state 5, and it is labeled an *undetected fault*. The 2λ failure rate is split between the two states per the two-outcome model, as shown in Fig. 3.2*b*. The undetected fault state is entered from the duplex state. If the fault is detected and corrected, the system will enter a simplex state without system failure; otherwise, the system will completely fail and enter failed state 3. Solution of this Markov model with real values is demonstrated in Chap. 4.

3.6.2 Markov Model of a Coverage Failure

Another peculiar type of failure occurs in a digital switching system when both subsystems are in working order and a switchover from the active subsystem to the standby subsystem is performed—the entire system fails. This type of switchover failure is called a *coverage failure*. The

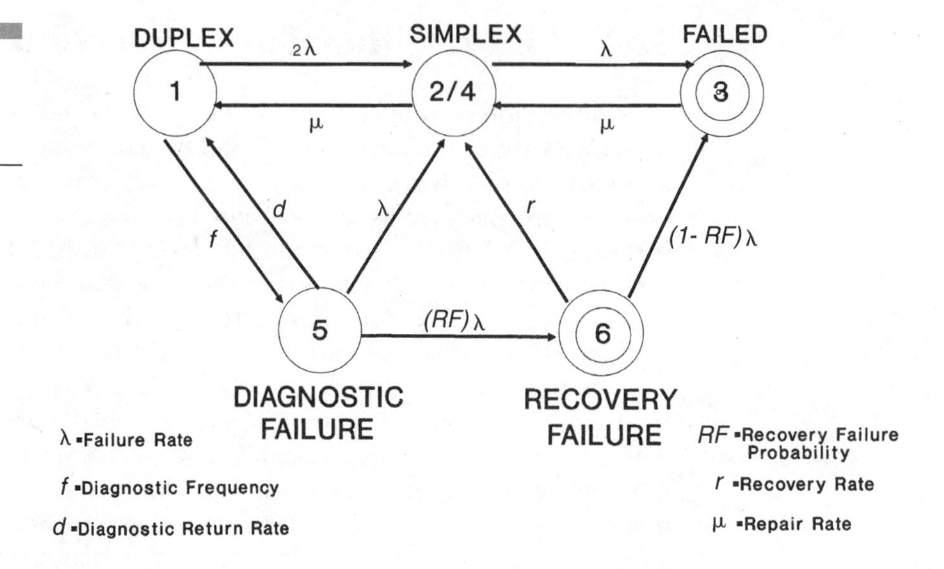

Figure 3.7 Diagnostic and recovery failure models

switchover may be actuated either automatically, due to some type of "hit" (a glitch in the power supply or data/address lines) or manually, during software/hardware maintenance process. A simplified Markov model of the coverage failure process is shown in Fig. 3.6*b*. Again, it is emphasized that the model depicted here shows a schema, and in complex systems the model could be far more elaborate. The figure shows transition state 5 as the coverage failed state. This state is entered from the duplex state with a coverage failure with probability CP. The coverage failure recovery rate is *R;* this is the rate at which the system is restored to a working simplex mode state 2/4. Note that state 5 in this case is a down state while in an undetected failure state 5 is not a down state. Solution of this Markov model with real values is demonstrated in Chap. 4.

3.6.3 Markov Model of Diagnostic and Recovery Failures

As another simple example of Markov modeling, consider the modeling of a diagnostic system of a digital switching system. Although the design specification for the diagnostic system requires that the diagnostic programs for digital switching systems find all faults in the system, in reality, this does not happen, and often the diagnostic programs cannot resolve all types of failure. The Markov model shown in Fig. 3.7 depicts such a system. Assume a digital switching system suffers a hardware failure. A diagnostic program is run to isolate the failing hardware. This program isolates the problem and takes the subsystem with a hardware fault out of service. The system suffers no downtime and is put in simplex mode (state 2/4). This diagnostic failure mode is represented by transition state 5. The diagnostic frequency rate is *f*, and the probability of successful completion of diagnostics is *d*. If the diagnostic program is unsuccessful in solving the problem or takes out a wrong piece of hardware from service, the system may be forced into a system recovery mode. *System recovery* is an emergency action needed to establish a working configuration for a digital switching system. This concept is discussed further in Chap. 7. During a recovery phase the digital switch is down. If the recovery process is successful, the system will be put into first a simplex configuration and then duplex mode. Transition state 6 in Fig. 3.7 shows the recovery failure mode. There is a finite probability that even the recovery program may fail to bring the system back to an operational mode. This type of probability is denoted RF (recovery failure probability). Under this condition the system will reach a permanent down state 3. Details of these concepts and their impact on system downtimes are given in Chap. 4.

3.6.4 Sensitivity Analysis

The term *sensitivity analysis* implies a methodology that studies the behavior of a system when the input to the system is varied. This technique can be applied to any modeling methodology. It allows different parameters of a model to be varied and the outcome to be recalculated. Different outcomes of modeling effort and corresponding simulated system behavior can then be studied by varying parameter settings.

In Markov models, this is best illustrated by an example. Consider the model of a duplicated subsystem in Fig. 3.4. The probability for state 1 is P_1, for state 2 is P_2, and for state 3 is P_3 (refer to Sec. 3.5.6 for details). Let us conduct three modeling runs, using three different values of repair strategy. In the first case, we assume the system has a failure rate of 10^{-5} failure per hour, and we wish to study the impact of varying repair times. Let us vary the repair rates from 2 to 5 h with increments of 1 h. By solving the transitional probabilities from the equation in Sec. 3.5.6, we find the following approximate results for P_1, P_2, and P_3:

Results for 2-h repair time:

Transition Phase	Calculated Value
P_1	0.9996
P_2	4×10^{-4}
P_3	8×10^{-8}

Results for 3-h repair time:

Transition Phase	Calculated Value
P_1	0.9994
P_2	6×10^{-4}
P_3	18×10^{-8}

Results for 4-h repair time:

Transition Phase	Calculated Value
P_1	0.9992
P_2	8×10^{-4}
P_3	32×10^{-8}

Results for 5-h repair time:

Transition Phase	Calculated Value
P_1	0.9990
P_2	9.9×10^{-4}
P_3	50×10^{-8}

In this particular case, the results can be easily interpreted. In a larger model, the effect of varying a large number of parameters may not be visualized so easily, and the results may have to be graphed or analyzed to better understand the impact of different parameters.

In this case, we see that the best strategy is to minimize the repair time, since each hour of delay in repairing the subsystem increases the unavailability of the system drastically. An increase of almost 125 percent occurs if the repair time is increased from 2 to 3 h, based on the values of P_3. The situation gets worse if the repair time is increased from 2 to 4 h—the unavailability of the subsystem increases by 300 percent. And for a repair time increase of 2 to 5 h, the unavailability becomes 525 percent higher. This is just a very simple example of how the parameters can be varied and of the impact of certain decisions on the reliability of a subsystem. In this case, clearly, faster repair times would make the subsystem more available, but without conducting a sensitivity analysis, it was not clear that the impact of repair time on availability would be so prominent.

3.7 Summary

This chapter introduced basic concepts related to the reliability analysis of digital switching systems. Reliability block diagrams, RPP, and Markov chain were introduced. Transition states and the solution of Markov chains using different methods were discussed. Some basic concepts related to the downtimes of a digital switching system such as coverage, undetected faults, diagnostic failures and system recovery. The reader was also introduced to the idea of sensitivity analysis.

REFERENCES

1. *Bellcore's Network Switching Outage Performance Monitoring Procedures,* SR-TSY-000963, Issue 1, April 1989.
2. Syed R. Ali, "Analysis of Outage Data for Stored Program Control Switching Systems," *IEEE Journal on Selected Areas in Communications,* vol. SAC-4, October 1986, pp. 1044–1046.
3. Harold Kushner, *Introduction to Stochastic Control,* Holt, Rinehart & Winston, New York, 1971.
4. John F. Kitchen, "Approximate Markov Modeling of High-Reliability Telecommunications Systems," *IEEE Journal on Selected Areas in Communications,* vol. SAC-4, no. 6, October 1986.
5. Tijms C. Henk, *Stochastic Models, An Algorithm Approach,* Wiley, New York, 1995, pp. 94–95, 368–369.

6. Daniel P. Heyman, "Further Comparisons of Direct Method for Computing Stationary Distributions of Markov Chains," *SIAM Journal on Algorithm and Discrete Mathematics,* vol. 8, no. 2, April 1987.
7. Winfried K. Grassmann, Michael I. Taksar, and Daniel P. Heyman, "Regenerative Analysis and Steady State Distributions for Markov Chains," *Operations Research,* vol. 33, no. 5, September–October 1985.
8. *Bellcore's Reliability Prediction Procedure for Electronic Equipment,* TR-332, Issue 5, December 1995.
9. *Bellcore's Methods and Procedures for System Reliability Analysis,* SR-TSY-001171, Issue 1, January 1989.

CHAPTER 4

Switching System Reliability Analysis

4.1 Introduction

The reader by now has acquired some basic working knowledge of a typical digital switching system and is also familiar with elements of Markov modeling. These two concepts are now combined and used to calculate and understand system downtimes of a digital switch.

4.2 Scope

This chapter employs reliability concepts developed in Chap. 3 to the digital switching systems developed in Chap. 2. To each digital switching system subsystem we apply Markov modeling, showing downtimes and pseudo-FIT rates. Various types of downtimes including system downtime are also covered.

4.3 Central Processor Community

Consider the central processor community that drives a digital switching system, developed in Chap. 2 (Fig. 2.1). The central processor portion of Fig. 2.1 is redrawn in Fig. 4.1*a*. A typical digital switching system's central processor unit (CPU) and its memory subsystem are duplicated or have a redundant architecture. For our purpose of developing basic concepts of reliability modeling, consider that the two central processors operate in parallel and that only one is active and controls the digital switching system while the other processor is fault-free and stays in a standby mode, ready to become active if the currently active processor fails.

4.3.1 State Transition Diagram

The state transition diagram is drawn to represent various operational and nonoperational states of the processor subsystem. Let us designate one of the central processors as side 0 and the other as side 1. If both sides are operational or "up," we call it a normal or *duplex* state and designate it as transition state 1. Designation of state numbers is arbitrary, but the type of state defined is important and should be noted on the diagram. Similarly,

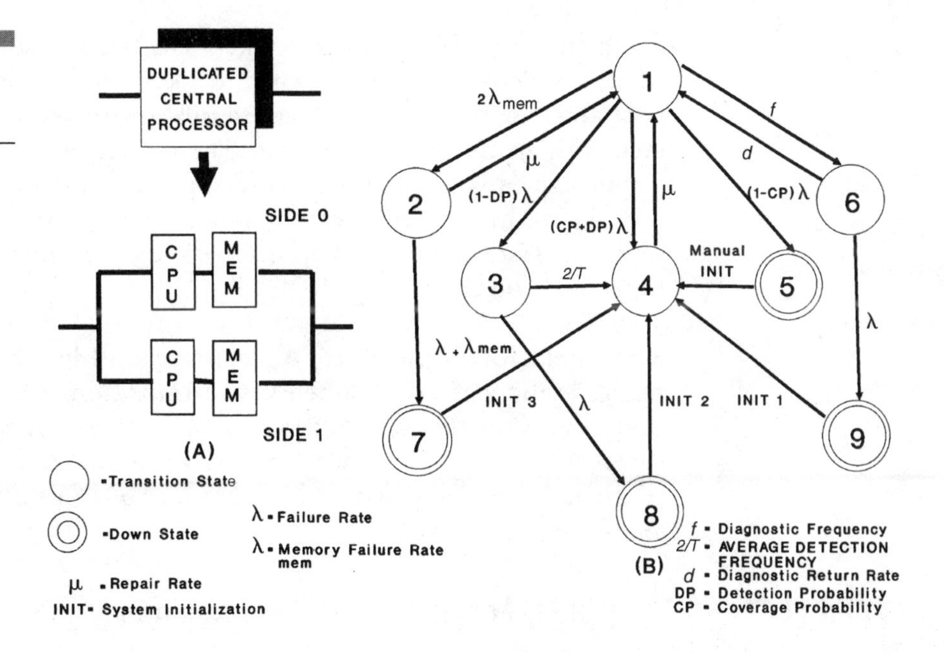

Figure 4.1 Markov model of a central processor

if either of the processors is not operational or is "down," a different state is designated, in this case as shown in Fig. 4.1*b*. States 2, 3, 4, and 6 represent *simplex* states. And if both the processors are down, then *failed* states exist and in this example are designated as states 5, 7, 8, and 9 and are enhanced by double circles. This, in essence, is the state transition diagram of the duplex central processor community.

Assumed Parameters The Markov model requires that all parameters be predefined. For sensitivity analysis, different parameters can be varied and the associated affects studied. To study the digital switching system developed in this book, we assume arbitrary failure-in-10^9-h (FIT) rates, coverage probability, detection probability, and other associated parameters. The recovery scheme for the central processor assumes a three-level recovery process. In this model, the lowest level of recovery is labeled INIT 1, and it takes 2 min to complete a low-level initialization process; this typically initializes certain predefined software registers and tables. The next level of initialization is denoted INIT 2, and it applies if INIT 1 cannot recover the system. The second-level initialization typically affects various translators and associated pointers. This type of initialization takes longer than INIT 1, and if it is unsuccessful, it is followed by the next higher level of initialization, labeled INIT 3 in this model. During INIT 3, all call paths are torn down, and all office parameters, translators, tables, pointers, etc., are initialized. In short, the whole system is reset or is returned to a known starting point. On some occasions, due to data or program mutilation, the system cannot recover even with an INIT 3 initialization. The last recourse under these severe circumstances is the manual recovery process, in which the entire software operating and application system is reinserted into the digital switching system from backup magnetic media. This process can sometimes take hours.

State Relationship Table A state relationship table is used to formally define various parameters and transition states of a subsystem under certain defined conditions. This table can typically be used to enter data into a Markov modeling program. Table 4.1 defines the parameter values and state transitions for the central processor subsystem shown in Fig. 4.1*b*.

TABLE 4.1

Modeling Results for Central Processor

Name of Model: Central Processor (CP)

Total number of states = 9

Total number of failed states = 4 (failed states 5, 7, 8, and 9)

Parameter or Value	Description
$\lambda = 250{,}000$	Failure rate of all components in FITs for the CP except memory subsystem
$\lambda_{mem} = 50{,}000$	Failure fate of the memory subsystem in FITs
$\mu = 1/3$	Repair rate, typically 3 h (1-h repair time, 2-h travel time)
$f = 1/24$	Diagnostic frequency (once in 24 h)
$d = 60/5$	Diagnostic return rate (average 5 min to complete diagnostic)
CP = 0.999	Coverage probability (typically 99.9%)
DP = 0.50	Detection probability by mate processor (typically 50%)
$T = 24$	Detection frequency once in 24 h, during the active/standby switchover at midnight
INIT 1 = 60/2	Low-level processor initialization (assumed to be 2 min)
INIT 2 = 60/10	Medium-level processor initialization (assumed to be 10 min, includes 2 min of INIT 1)

TABLE 4.1

Continued

INIT 3 = 60/20	High-level processor initialization (assumed to be 20 min, includes 10 min of INIT 2)
Manual INIT = 60/60	Time required for manual processor initialization (assumed to be 60 min)

Transitions

From State	To State	Equation	Reason
1	2	$2\lambda_{mem}$	Transition to simplex mode due to memory failure
1	3	$(1 - DP)\lambda$	Transition to simplex mode due to fault that could not be detected in standby processor
1	4	$(CP + DP)\lambda$	Transition to simplex mode due to fault in either processor (standby or active)
1	5	$(1 - CP)\lambda$	Transition to down state due to fault uncovered in active processor
1	6	f	Transition to simplex mode for midnight diagnostics
2	1	μ	Transition back to duplex state after memory repair
2	7	$\lambda + \lambda_{mem}$	Transition to down state if fault occurs when processor is in simplex mode due to memory problems
3	4	$2/T$	Transition to simplex state due to midnight switchover (mean time)
3	8	λ	Transition to down state due to failure in active processor
4	1	μ	Transition to duplex state after processor repair
5	4	M INIT	Transition from down state to simplex state brought about by manual initialization
6	1	d	Transition to duplex state after successful diagnostic
6	9	λ	Transition to down state if failure occurs while system is in simplex mode for diagnostics
7	4	INIT 3	Transition from down state to simplex state due to level 3 initialization

TABLE 4.1 Continued

8	4	INIT 2	Transition from down state to simplex mode due to level 2 initialization
9	4	INIT 1	Transition from down state to simplex mode due to level 1 initialization

Availability	0.99999
Downtime	0.19380 min/yr

4.4 Clock Subsystem

One of the major causes of duplex failures in digital switches is attributable to clock subsystem failures. Modern digital switches always employ redundant clock systems. Failure of clocks can cause memory mutilation, time slot misassignments, interface timing problems with other subsystems, synchronization difficulties with other switches, and myriad other troubles. This is one reason why the clock subsystem was chosen for modeling.

4.4.1 State Transition Diagram

The state transition diagram for the clock subsystem is shown in Fig. 4.2. It follows the same basic definitions previously described for the

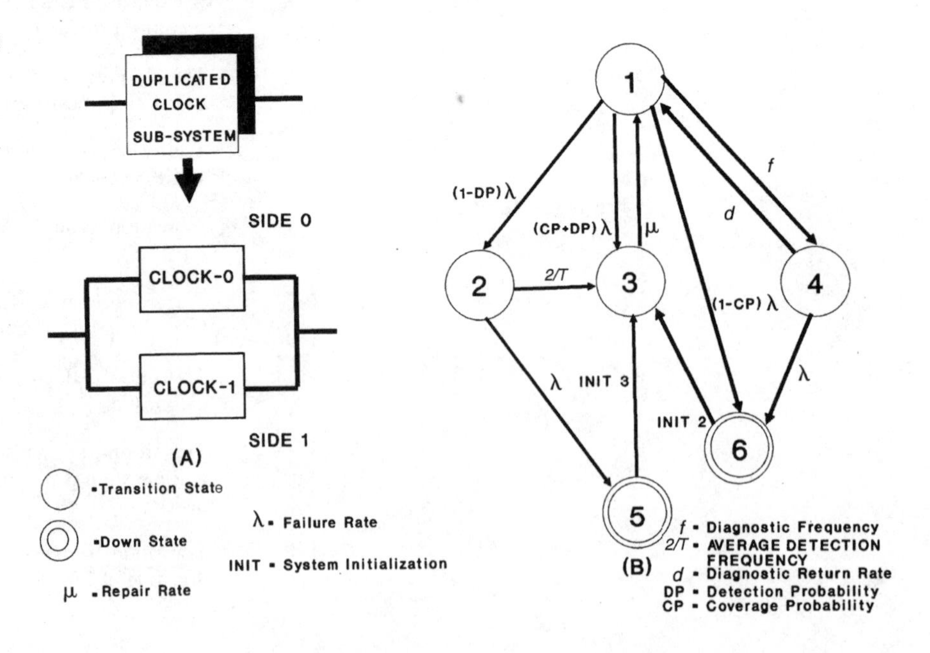

Figure 4.2 Markov model of a clock subsystem

processor community. Again the parameters are arbitrarily chosen. It is also assumed that during the initialization caused by clock subsystem failure, INIT 1 will not be sufficient to effect system recovery, and a higher level of initialization will be required. INIT 2 will recover the system during diagnostic or coverage failures. However, if the system reached the down state due to detection problems, then it is assumed that the problem is not easily diagnosable and INIT 3 initialization will be necessary for system recovery. Again it must be emphasized that this is just a model contrived by the author; different clock systems may have different mechanisms for system recovery, etc.

State Relationship Table Table 4.2 defines the parameter values and state transitions for the clock subsystem shown in Fig. 4.2*b*.

TABLE 4.2

Modeling Results for Clock Subsystem

Name of Model: Clock Subsystem (CLK)

Total number of states = 6

Total number of failed states = 2 (failed states 5, 6)

Parameter or Value	Description
$\lambda = 50{,}000$	Failure rate of all components in FITs for clock subsystem
$\mu = 1/3$	Repair rate, typically 3 h (1-h repair time, 2-h travel time)
$f = 1/24$	Diagnostic frequency (once in 24 h)
$d = 60/5$	Diagnostic return rate (average 5 min to complete the diagnostic)
CP = 0.999	Coverage probability (typically 99.9%)
DP = 0.50	Detection probability by mate processor (typically 50%)
$T = 24$	Detection frequency once in 24 h, during active/standby switchover at midnight
INIT 2 = 60/10	Medium-level processor initialization (assumed to be 10 min, includes 2 min of INIT 1)
INIT 3 = 60/20	High-level processor initialization (assumed to be 20 min, includes 10 min of INIT 2)

TABLE 4.2 Continued

Transitions

From State	To State	Equation	Reason
1	2	$(1 - DP)\lambda$	Transition to simplex mode due to clock fault that could not be detected in standby processor
1	3	$(CP + DP)\lambda$	Transition to simplex mode due to clock fault in either processor
1	4	f	Transition to simplex mode for midnight diagnostics
1	6	$(1 - CP)\lambda$	Transition to down state due to clock fault uncovered in active processor
2	3	$2/T$	Transition to simplex mode due to midnight switchover
2	5	λ	Transition to down state due to clock failure in active processor
3	1	μ	Transition to duplex mode after processor repair
4	1	d	Transition to duplex mode after successful diagnostic
4	6	λ	Transition to down state if failure occurs while system is in simplex mode for diagnostics
5	3	INIT 3	Transition from down state to simplex mode due to level 3 initialization
6	3	INIT 2	Transition from down state to simplex mode due to level 2 initialization

Availability	0.99999
Downtime	0.02212 min/yr

4.5 Network Controller Subsystem

The next subsystem to model is the network controller subsystem, and it was chosen to demonstrate modeling techniques that could be applied to subsystems that are mostly autonomous and are usually not dependent on central processor (CP) control. This model assumes that 80 percent of the time the network controller can be repaired without any assistance from the CP. However, the other 20 percent of the time, the CP is involved and causes a midlevel recovery action. Again, this is just a model, and various network controllers may have entirely different recovery strategies. Note that this model does not use midnight diagnostics and other failure modes associated with coverage and detection.

From the modeling results it is evident that the contribution to the total system downtime is minimal, which is attributable to its autonomous recovery scheme.

4.5.1 State Transition Diagram

The state transition diagram for the network controller shows that there is only one down state. The network controller may suffer a duplex failure, as shown in the transition from state 1 to state 2 and then to state 3. Normally, a duplex controller does not cause the entire system to go down, but only impacts those lines that are assigned to the controller subsystem, therefore causing only a partial outage. However, if the duplex controller failure is such that it requires system initialization and recovery, then the system downtime will be impacted as depicted in state 4.

State Relationship Table Table 4.3 defines the parameter values and state transitions for the network controller shown in Fig. 4.3*b*.

TABLE 4.3

Modeling Results for Network Controller

Name of Model: Network Controller (CONT)

Total number of states = 4

Total number of failed states = 2 (failed states 3, 4)

Parameter/Value	Description
$\lambda = 130{,}000$	Failure rate of all components in FITs for network controller
$\mu = 1/3$	Repair rate, typically 3 h (1-h repair time, 2-h travel time)
$r = 60/5$	Diagnostic return rate for the controller (average 5 min to complete the diagnostic)
RF = 0.80	Independent recovery probability (assumed 80 percent)
INIT 2 = 60/10	Medium-level processor initialization (assumed to be 10 min, includes 2 min of INIT 1)

TABLE 4.3

Continued

Transitions

From State	To State	Equation	Reason
1	2	2λ	Transition to simplex mode due to network controller fault
2	1	μ	Transition to duplex mode after controller repair
2	3	$RF \cdot \lambda$	Transition to down state due to duplex controller failure (independent recovery)
2	4	$(1 - RF)\lambda$	Transition to down state due to duplex controller failure (CP-assisted recovery)
3	2	r	Independent controller recovery rate (assumed 5 min)
4	2	INIT 2	Transition from down state to simplex mode due to level 2 initialization, required for recovery via CP

Availability	0.99999
Downtime	0.00533 min/yr

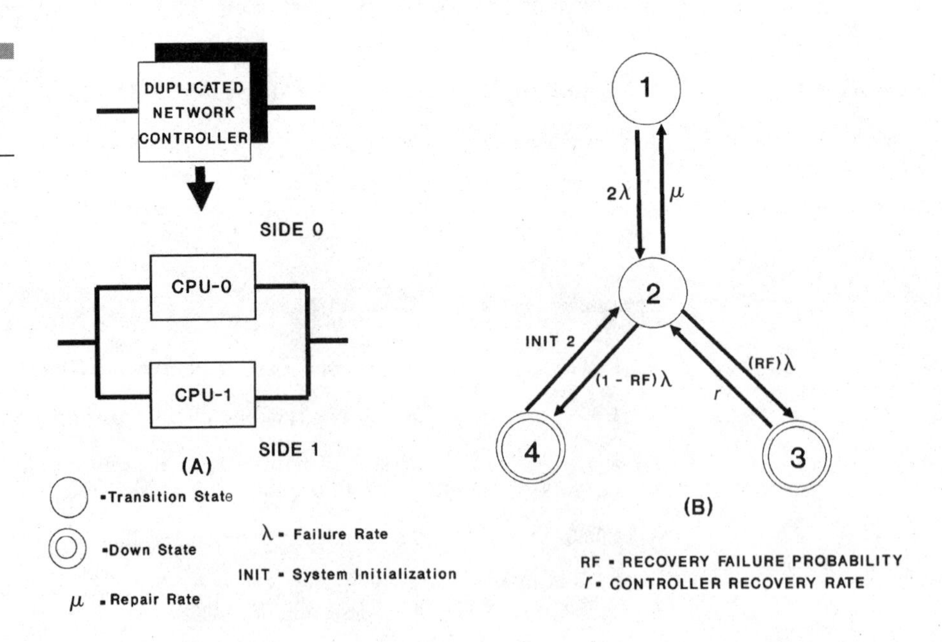

Figure 4.3 Markov model of a network controller

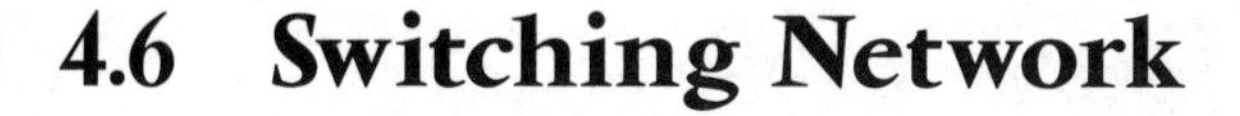

4.6 Switching Network

The Markov model presented here assumes that the switching network is duplicated in two planes, plane 0 and plane 1. All network orders (connecting, tear-down, maintenance, etc.) are simultaneously acted upon on both planes. In the case of a severe network problem, the network controller has the capability of switching the network planes without losing any call connection. Again, this is just a model developed for illustration purposes, and other means could be used to provide redundancy. Refer to Chap. 2 for more details on various types of switching fabric and how they are controlled.

4.6.1 State Transition Diagram

The Markov model for this switching network assumes a six-state model. The usual states describing midnight diagnostics, coverage, and detection states are the same as in other models. It is also assumed that the switching network will be able to switch planes automatically without incurring any downtimes. However, the switching network can enter duplex failure mode due to coverage failure, and it is assumed that 75 percent of the time, the network controller will be able to recover without any assistance from the central processor (CP). The other 25 percent of the time, the CP will be required to start a medium-level initialization process (INIT 2) to complete the switching network recovery process.

State Relationship Table Table 4.4 defines the parameter values and state transitions for the switching network subsystem shown in Fig. 4.4*b*.

TABLE 4.4

Modeling Results of Switching Network

Name of Model: Switching Network (NTK)

Total number of states = 6

Total number of failed states = 2 (failed states 5, 6)

Parameter/Value	Description
$\lambda = 200{,}000$	Failure rate of all components in FITs for switching network
$\mu = 1/3$	Repair rate, typically 3 h (1-h repair time, 2-h travel time)
$f = 1/24$	Diagnostic frequency (once in 24 h)

TABLE 4.4
Continued

Parameter/Value	Description
$d = 60/5$	Diagnostic return rate (average 5 min to complete the diagnostic)
CP = 0.999	Coverage probability (typically 99.9%)
DP = 0.50	Detection probability by mate processor (typically 50%)
$T = 24$	Detection frequency once in 24 h, during active/standby switchover at midnight
NR = 60/3	Network plane switchover or reconfiguration time (3 min)
INIT 2 = 60/10	Medium-level processor initialization (assumed to be 10 min, includes 3 min of INIT 1)
$P = 0.75$	Probability that network will reconfigure with CP initialization

Transitions

From State	To State	Equation	Reason
1	2	$(1 - \text{DP})\lambda$	Transition to simplex mode due to network fault
1	3	$(\text{CP} + \text{DP})\lambda$	Transition to simplex mode due to network fault in either plane
1	4	f	Transition to simplex mode for midnight diagnostics
1	6	$(1 - \text{CP})\lambda$	Transition to down state due to network fault uncovered in active plane
2	3	$\frac{2}{T}$	Transition to simplex mode due to midnight switchover
3	1	μ	Transition to duplex state after network repair
4	1	d	Transition to duplex state after successful diagnostic
4	6	λ	Transition to down state if failure occurs while system is in simplex mode for diagnostics
5	3	INIT 2	Transition from down state to simplex mode due to level 2 initialization
6	3	$(P) \cdot \text{NR}$	Network recovery without CP initialization
6	5	$(1 - P)\text{NR}$	Network recovery with CP initialization

Availability	0.99999
Downtime	0.04284 min/yr

Figure 4.4
Markov model of a switching network

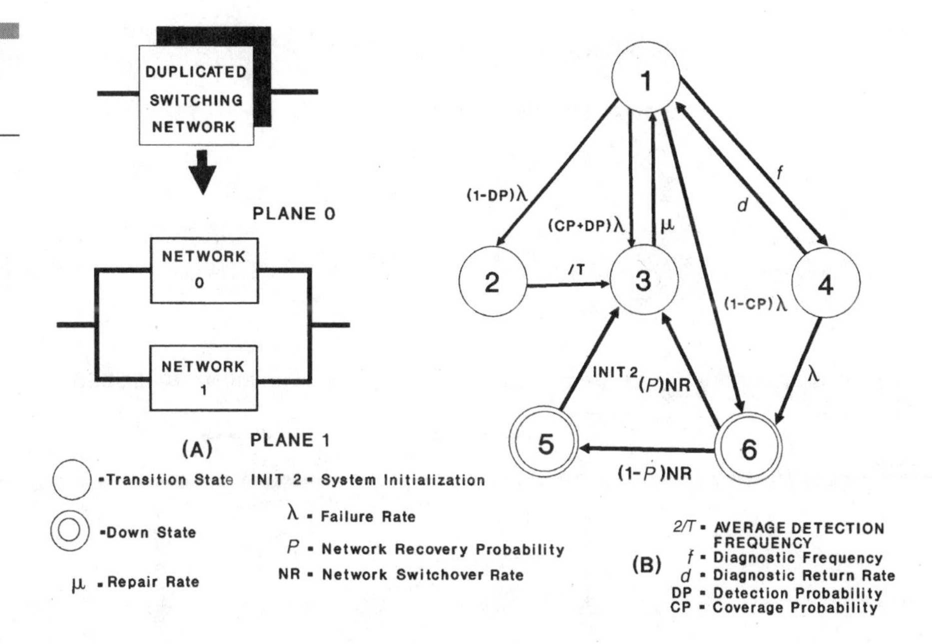

4.7 System Downtimes Due to All Hardware Failures

All the major components of the digital switching system developed so far have been modeled. The only components that remain are the line and trunk modules. These modules do not ordinarily impact system downtimes and therefore are treated separately.

Table 4.5 adds all the contributions to the downtimes of the Markov model in previous sections. The results show that the digital switching system developed in this book will suffer a downtime of 0.26 min/yr due to hardware failures alone.

4.8 Line and Trunk Downtimes

The Markov models for lines and trunks are identical but usually have different failure rates. Since the line equipment is not duplicated, its Markov model is very simple and usually does not require more than three states. If the digital switch supports remote lines, then the model developed below assumes 1 h extra of travel time. This in effect increases the downtimes associated with the remote lines.

TABLE 4.5
Overall Modeling Result

Subsystem	Downtime, min/yr
Central processor	0.194
Clock subsystem	0.022
Network controller	0.005
Switching network	0.043
Total	0.264

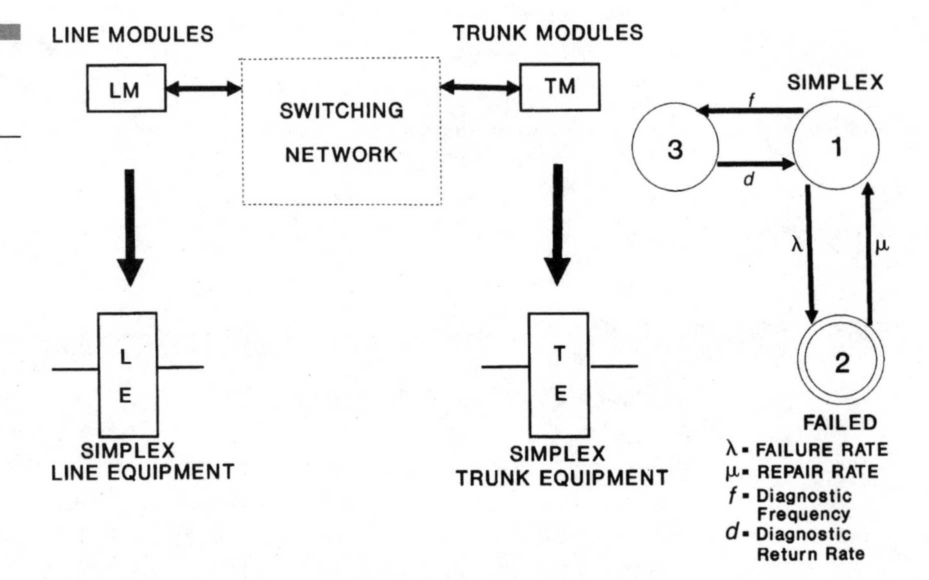

Figure 4.5
Markov model of a line or a trunk

4.8.1 State Transition Diagram

The state transition diagram for the lines and the trunks is shown in Fig. 4.5. The lines and trunks operate in simplex mode, state 1. In case of line or trunk equipment problems, they enter a down state, state 2. At midnight, most of the digital switching systems conduct a routine test on lines or trunks, shown as state 3. Although the lines or trunks are busy during this test and are not accessible to customers, technically, they are not considered to be down.

State Relationship Table Table 4.6 defines the parameter values and state transitions for the lines and trunks shown in Fig. 4.5. Also see Table 4.7 for modeling results for lines and trunks.

4.9 Call Cutoffs

Practically all telephone users, at one time or another, have experienced a call cutoff. This situation occurs when users get cut off through no action of their own, i.e., hanging up or flashing. These call cutoffs primarily result from hardware problems in a digital switching system.

TABLE 4.6

Modeling Results for Lines and Trunks

Name of Model: Line or Trunk

Total number of states = 3

Total number of failed states = 1 (failed state 2)

Parameter or Value	Description
$\lambda = 15{,}000$	Failure rate of all components in FITs for lines
$\lambda = 10{,}000$	Failure rate of all components in FITs for trunks
$\mu = 1/3$	Repair rate, typically 3 h for regular lines/trunks (1-h repair time, 2-h travel time)
$\mu = 1/4$	Repair rate, typically 4 h for remote lines (1-h repair time, 3-h travel time for remote lines)
$f = 1/24$	Diagnostic frequency (once in 24 h)
$d = 60/1$	Diagnostic return rate for lines (assumed to be 1 min)
$d = 60/2$	Diagnostic return rate for trunks (assumed to be 2 min)

Transitions

From State	To State	Equation	Reason
1	2	λ	Transition to down state due to line/trunk fault
1	3	f	Diagnostic frequency (once in 24 h)
2	1	μ	Transition to simplex mode after line/trunk repair
3	1	d	Diagnostic return rate

TABLE 4.7

Modeling Results for Lines and Trunks—Summary

Line or Trunk	FIT Rate	Downtime, min/yr
Regular lines	15,000	23.63
Remote lines	15,000	31.51
Regular trunks	10,000	15.75

They are also due to software problems or other problems unrelated to the action of the user. An excellent paper by Tortorella[1] describes call cutoffs in detail. Bellcore[2] also describes more accurate ways of calculating call cutoffs and other related measurements. The probability of call cutoffs can be approximated by

$$\text{Cutoff prob.} = \sum_{i=1}^{n} \lambda(i)\,\text{CF}(i) C_{\text{dur}}$$

where i = failure mode number
n = total number of failure modes
$\lambda(i)$ = failure rate in FITs for failure mode i
CF(i) = Connection factor (defines type of connection) can be assumed to be 1 for total line or total trunk connection losses; otherwise, 2 can be assumed
C_{dur} = mean call duration, h

The values of CF(i) need to be determined for each mode i. For a more detailed treatment of this subject, consult Ref. 2, section 12.

EXAMPLE 4.1 Assume that a digital switching system is equipped with 30,000 lines and 4000 trunks and exhibits four failure modes, or types of failures. In the first mode, all 30,000 lines and 4000 trunks are lost; in the second mode, 3000 lines and 200 trunks are lost; in the third mode, 300 lines and 20 trunks are lost; and in the fourth mode, 30 lines and 5 trunks are lost. Also assume that the duration of calls is about 8 min (0.13 h) and FIT (1 failure in 10^9 h = 1 FIT) rates of 5000, 700, 90, and 25, respectively, for each mode of failure.

An approximate way to calculate the call cutoffs for this central office is shown in Table 4.8, where the approximate cutoff probability equation is evaluated for each failure mode and then added to obtain the result.

TABLE 4.8

Approximate Call Cutoff Probability

Mode	No. Lines	No. Trunks	$\lambda(i)$ FITs	CF(i)	C_{dur}	Cutoff Prob. $\times 10^{-9}$
1	30,000	4,000	5,000	1	0.13	650
2	3,000	200	700	2	0.13	182
3	300	0	90	2	0.13	23.4
4	0	5	25	1	0.13	3.3

The sum of all four modes yields a probability cutoff of approximately 859×10^{-9}, which implies that about 860 out of 1 billion calls will experience call cutoffs.

4.10 Ineffective Machine Attempt

Sometimes a telephone user may not be able to complete a call due to unavailability of central office (CO) equipment. For instance, a customer may not be able to initiate a call during a busy hour, since all the equipment that can provide dial tones to customer lines may be in use. This type of failure is termed *reliability-related ineffective machine attempt.* The probability of this occurring is termed the *ineffective machine attempt* (*IMA*).

It is important to distinguish this type of failure mode from call cutoffs, and vice versa. For instance, the switching network in Fig. 3.9 may be in the process of being switched from one plane to another. Some calls may be dropped during this switchover action. This type of failure should be classified as call cutoff, not IMA. Currently, the concept of IMA is supported for hardware-type failures only. The IMA can be approximated by

$$\text{IMA prob.} = \sum_{i=1}^{n} \text{CF}(i) \cdot \text{SU}(i)$$

where i = failure mode number
n = total number of failure modes
CF(i) = connection factor, defines type of connection, can be assumed to be 1 for Total Line or Total trunk connection losses, otherwise an approximate value of 2 may be used.
SU(i) = system unavailability for failure mode i

The system unavailability can be calculated from

$$\text{SU} = 1 - A$$

where A is the availability of the system, which in turn can be obtained from

$$A = \frac{\text{MTBF}}{\text{MTBF} + \text{MTTR}}$$

where MTBF = mean time before failure and MTTR = mean time to repair. Also

$$\text{MTTR} = \frac{1 - A}{\lambda} \qquad \text{or} \qquad \text{SU} = \lambda \bullet \text{MTTR}$$

EXAMPLE 4.2 To extend Example 4.1 for the calculation of IMA, we need the system availability for each of the four failure modes, assuming that these modes can cause IMA. It can be easily calculated by multiplying the failure rates of Example 4.1 by their respective repair rates. Assume a repair rate of 6 h for the first mode, 3 h for the second mode, and 2 h for the last two modes. We then obtain the approximate IMA in Table 4.9 by simple multiplication. The sum of all four modes yields a probability of IMA of $34{,}430 \times 10^{-9}$, which implies that about 34,000 out of 1 billion calls will experience ineffective machine attempts.

4.11 Partial Downtimes

As discussed earlier, currently most digital switching systems employ distributed processing. One by-product of this technology is distributed outages, and it can be said that *with distributed processing come distributed problems!* Partial downtimes in the first generation of digital switching systems were practically unknown. When the system went down, all lines and trunks were lost; and when the system came back up, they all came back up. Not so in today's digital switch environment. Earlier systems used a common control architecture; today, most of digital switches use various hierarchies of controls, as discussed in Chap. 2. Depending on the architecture, various groups of lines and trunks may go out of

TABLE 4.9

Approximate IMA Probability

Mode	No. Lines	No. Trunks	$\lambda(i)$ FITs	MTTR, h	CF(i)	SU(i)	IMA$\times 10^{-9}$
1	30,000	4,000	5,000	6	1	30,000	30,000
2	3,000	200	700	3	2	2,100	4,200
3	300	0	90	2	1	180	180
4	0	5	25	2	1	50	50

service, while the majority of the system may still be up and running. This type of outage is classified as a *partial outage* and it can be calculated easily by

$$\text{Partial downtime} = \frac{\sum_{i=1}^{n} \text{DT}(i)\text{PTL}(i)}{\text{TL}}$$

where i = failure mode number
n = total number of failure modes
$\text{DT}(i)$ = downtime associated with failure mode i
$\text{PTL}(i)$ = number of trunks and/or lines affected during partial outage for mode i
TL = total number of lines and/or trunks in system

EXAMPLE 4.3 Building further on the previous examples of call cutoffs and IMA and assuming the same parameters, we can calculate the associated partial downtimes. Obviously mode 1 will not qualify for partial outage since all the lines and trunks are lost in that mode. The other three modes can cause partial outages. Assuming downtimes of 10, 30, and 60 min for the last three modes, respectively, we obtain

$$\text{Partial downtime mode 1} = 0$$

$$\text{Partial downtime mode 2} = 10\left(\frac{3{,}200}{34{,}000}\right) = 0.94 \text{ min}$$

$$\text{Partial downtime mode 3} = 30\left(\frac{300}{34{,}000}\right) = 0.26 \text{ min}$$

$$\text{Partial downtime mode 4} = 60\left(\frac{5}{34{,}000}\right) = 0.01 \text{ min}$$

$$\text{Partial downtime for all modes} = 0.94 + 0.26 + 0.01 = 1.21 \text{ min}$$

The concept of partial downtimes can be extended to remote line and trunk units as well. The table below summarizes the results of the last three examples.

Cause	Frequency
Call cutoffs	860 calls/1 billion calls
Ineffective machine attempts	34,000 calls/1 billion calls
Partial downtime	1.21 min

4.12 Summary

This chapter has shown how Markov modeling methods can be practically utilized to predict the downtimes of a digital switching system and how line and trunk downtimes can be predicted if accurate failure rates of lines and trunks are available. This chapter discussed call cutoffs, ineffective machine attempts, and partial downtimes.

REFERENCES

1. M. Tortorella, "Cutoff Calls and Telephone Equipment Reliability," *Bell Systems Technical Journal,* vol. 60, no. 8, October 1981.
2. *Bellcore, LATA Switching Systems Generic Requirements, Reliability,* Section 12, TR-TSY-00512, Issue 3, February 1990.

CHAPTER 5

Switching System Software

5.1 Introduction

Clearly, a modern digital switching system is quite complex. At this stage of digital switching evolution, most of the complexity comes from the software, which is much more complex and harder to manage than the hardware it controls. This chapter exposes some intricacies of the software that drives a digital switching system. The techniques for analyzing switching system software are covered in Chap. 6.

5.2 Scope

This chapter covers the basic software architecture of a typical digital switch, classifies various types of software, describes a basic call model and software linkages that are required during a call, and discusses some basic call features.

5.3 Basic Software Architecture

A good understanding of the hierarchy of software currently employed by many modern digital switching systems is important. Most of today's digital switching systems employ quasi-distributed hardware and software architectures. The control structure of a digital switching system can usually be divided into three distinct levels, as described in Chap. 2. This chapter elaborates on the software employed in a hypothetical digital switching system at different levels of control. A modified version of Fig. 2.1 is shown here as Fig. 5.1. It shows levels of control along with some details of minimum software architectural functions that may be necessary for each level of control. From a digital switching system analyst's point of view, it is essential to understand the high-level software architecture of a digital switch before attempting to analyze it. Low-level details are not essential, since the objective is to analyze the digital switch, not to design it. However, details on software engineering practices are essential for software analysis. This is covered in Chap. 6. The next three sections identify software components that are usually necessary for a modern digital switching system.

5.3.1 Operating Systems

Every digital switching system has an operating system as a part of its software architecture. An *operating system* (*OS*) may be defined as software that manages the resources of a computer system or controls and tasks other programs. Sometimes these programs may be referred[1] to as control programs, supervisory programs, executive programs, or monitor programs. In theory, there are different types of operating systems, classified as serial batch systems, multiprogramming systems, timesharing systems, and the real-time systems.[2] The operating systems employed by digital switching systems are real-time operating systems. This type of OS is required for digital switching systems since the very nature of telephony processing demands execution of tasks in real time. Typically, the real-time operating system for the digital switching system interacts with different layers of applications necessary to support telephony features and functions. Since practically all modern digital switching systems use quasi-distributed architecture, the processor or controller for each subsystem may even use different OSs than the central processor does.

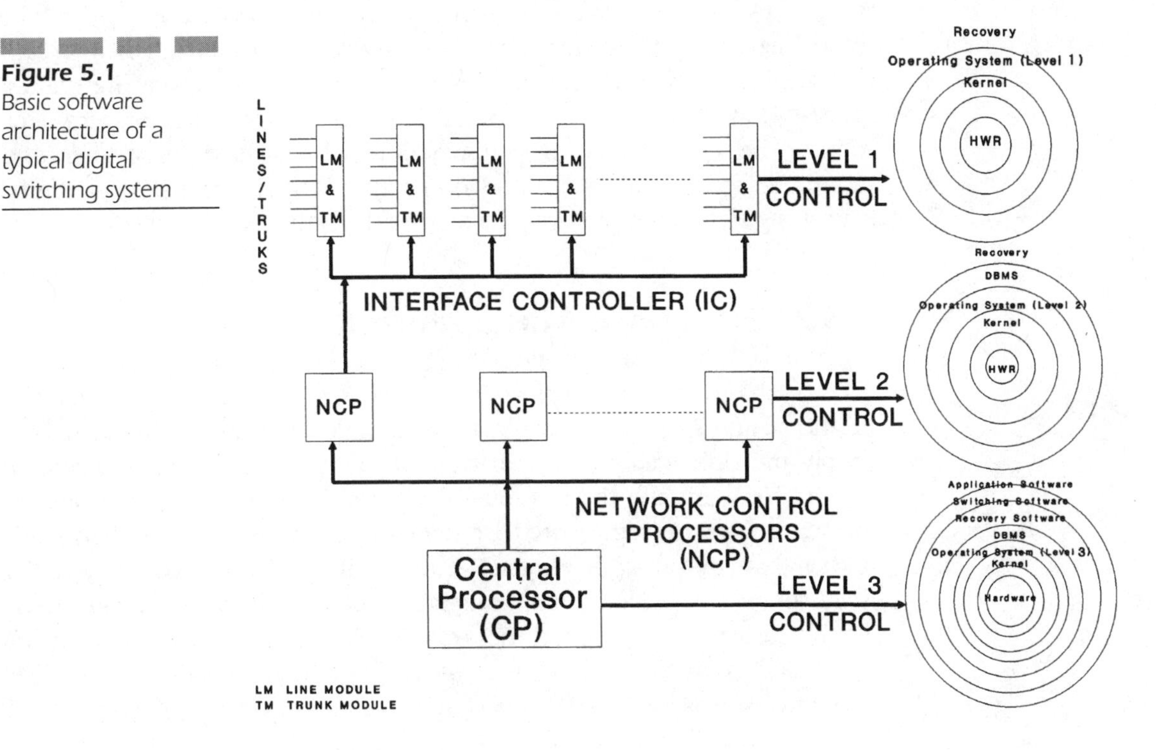

Figure 5.1 Basic software architecture of a typical digital switching system

Therefore, it is conceivable for a digital switching system to employ more than one OS. Each subsystem may employ a different type of processor and therefore may employ different high-level languages for the development of its software. It is thus a challenge for the analyst to understand the operational and developmental environment of a digital switching system. Details on a methodology that could aid the analyst in better understanding the complex software environment of digital switching systems are given in Chap. 6.

Kernel The kernel or the nucleus of an operating system comprises those functions of an OS that are most primitive to the environment. It usually supports the following functions.[2]

- Process control and scheduling
- Main memory management
- Input/output control of requests for terminals and buffers
- Domain protection of main memory read/write operations etc.

Most of the real-time operating systems that control digital switching systems use priority interrupt systems. These interrupts are serviced by the kernel based on the importance of an operation. Of course, the interrupt and similar hierarchical controls are system-specific, but most give highest priority to system maintenance interrupts, since this ensures proper operation of the digital switching system, followed by other types of interrupts required for call processing and other ancillary functions. Most digital switching systems employ kernels that reside in the main memory.

5.3.2 Database Management

The databases that are employed in digital switching systems are usually relational and sometimes distributed. In simple terms, distributed databases imply multiple databases requiring data synchronization. The relational database systems use the relational data model in which the relationships between files are represented by data values in file records themselves rather than by physical pointers.[3] A record in a relational database is flat, i.e., a simple two-dimensional arrangement of data elements. The grouping of related data items is sometimes referred to as a *tuple*. A tuple containing two values is called a pair. A tuple containing *N* values is called an *N*-tuple.[4] A good example of a relational database system in a digital switching system is a database system that keeps cross-references of all directory numbers that are assigned to the line equipment of subscribers. When a particular subscriber goes off-hook, the line equipment is identified by the scanning program. The database is searched to find its associated directory number that identifies all characteristics of the line. In the hypothetical digital switching system developed for this book, each network control processor (NCP) is assigned a group of subscribers. Therefore, each NCP has a replica of the subscriber database for all other NCPs. Depending on the type of call, a NCP may be required to route calls through other NCPs. To accomplish this, the database information for all NCPs needs to be distributed and always kept synchronized.

5.3.3 Concept of Generic Program

Most digital switching systems support the concept of *generic program* similar to *release* in the computer industry. However, a generic program can be a little more involved than release. In the early days of stored program control (SPC) switching systems, the generic program was more or less the same for most telephone companies. Usually, the generic program contained all programs necessary for the switching system to function. It included all switching software, maintenance software, and specialized office data for the configuration of a central office (CO). The translation data were usually supplied by the telephone companies. However, now it is sometimes more difficult to define exactly what a generic program consists of. Most of the modern digital systems have modular software structure. They usually have *base* programs or *core* programs that control the basic functions of the digital switching systems. On top of these programs reside different features and special options. Generally, the performance of a generic program is tracked for software reliability. Therefore, it becomes very important for an analyst to identify the exact software components that constitute a generic program. The components of a digital switching system's software that are kept common for a specific market or a group of telephone companies can sometimes be used to identify the generic program. Usually, this set of programs can be labeled as a generic, base, or core release for a digital switching system. In general, generic programs contain operating system(s), common switching software, system maintenance software, and common database(s) software for office data and translation data management.

5.3.4 Software Architecture for Level 1 Control

Level 1 is the lowest level of control. This level is usually associated with lines, trunks, or other low-level functions. Most of the software at this level is part of the switching software. As shown in Fig. 5.1, the interface controllers (ICs) are usually controlled by microprocessors and may have a small kernel controlling the hardware of the IC. The ICs may have a small OS, labeled Operating System (Level 3) in Fig. 5.1. The function of this OS is to control and schedule all programs that are resident in the IC. Most of the ICs have enough intelligence to recognize proper functioning of hardware and software. The IC can also conduct diagnostics of lines and trunks or other peripherals connected to it. More extensive diagnostic routines may reside in the central processor or in some cases in the IC itself. In either case, the central processor can run the diagnostic program itself or request a fault-free IC to run it. The IC will then run the diagnostics and forward the results to the central processor. The ICs may also be capable of *local* recovery. This means that in case of an IC failure, the IC could recover itself without affecting the entire digital switching system. The only effect will be on the lines and trunk or peripherals connected to the IC undergoing a recovery process. Again, all this will depend on the design of the ICs and associated software. An analyst should be conversant with different types of design strategies that may be employed, since they will impact the reliability and functionality of the IC.

5.3.5 Software Architecture for Level 2 Control

The intermediate or level 2 controls are usually associated with network controllers that may contain distributed databases, customer data, and service circuit routines. Obviously, these functions are digital switching architecture-dependent; many switching functions could be assigned at this level of control. In a quasi-distributed environment, the processors employed are usually of intermediate or mini size. The NCPs are usually independent of the central processor. As shown in Fig. 5.1, the NCPs usually have their own operating system, labeled Operating System (Level 2). This OS has a kernel that controls the hardware and basic functionalities of the NCP. At this level of control, usually a resident database system maintains the translation data of subscribers and other software parameters required to control the telephony functions of the NCP. System recovery at this level of control is crucial, since a failure of a NCP may impact a number of ICs (dependent on the design) and a large number of lines, trunks, and peripherals. The NCPs should be capable of self-diagnosis, and since they are duplicated, they must be able to switch to a working backup. As mentioned in earlier chapters, the use of NCPs is design-specific. A design may call for a dedicated NCP to act as a control NCP for all other NCPs, or each NCP may be designed to operate independently. The recovery strategy in each case will be different. In the first case, where one NCP acts as the control NCP, the control NCP is responsible for system recovery for all other NCPs. In the second case, where there is no control NCP, the central processor is responsible for the recovery process of all NCPs. There could be all kinds of recovery strategies involved in the system recovery process at this level. The analyst needs to understand what type of recovery strategy is being used, in order to better assess the reliability of a digital switching system.

Consider the function of the NCP. A subscriber goes off-hook, the IC receives an off-hook notification from the line module. The IC requests details on the subscriber, such as allowed features and applicable restrictions. The NCP queries its database for this information and passes it back to the IC. This type of action required by the NCP necessitates that the NCP maintain a subscriber database as well. This database is supposed to be managed and kept up to date with the latest information for each subscriber. This is shown as DBMS in Fig. 5.1 under level 2 control. A detailed explanation of how this type of database is utilized for routing calls in a hypothetical digital switching system is given in Chap. 9.

5.3.6 Software Architecture for Level 3 Control

The highest or level 3 control is usually associated with the central processor of a digital switching system. Normally these processors are mainframe-type computers. Usually, the CP of a digital switching system provides all high-level functions. These high-level functions include the management of the database system for office data, high-level subscriber data, software patch levels, feature control, and above all, system recovery in case of hardware or software failures. The main operating system of a modern digital switching system resides at this level and is labeled Operating System (Level 3) in Fig. 5.1. As mentioned earlier, this OS operates in real time and is multitasking (i.e., it can support more than one task at a time). This OS controls the database management system, switching software, recovery software, and all applications such as features, traffic management systems, and OS interfaces. Most CPs work in an active/standby mode. In this mode, one CP is always available to go into active mode if the active CP develops a fault. Indeed, there are different schemes for operating a redundant processor system to improve reliability and availability. However, for digital switching systems, the scheme most commonly employed is the one in which both processors execute instructions in a matched mode, and in case of a failure, the standby processor becomes active immediately. Other schemes are sometimes employed, such as hot standby, in which the standby processor is powered up and ready to take over the operation of an active processor. In this scheme, call processing can be impacted during the processor switchover. There is a third option, cold standby, in which the processor is not powered up, but can be brought on line in case of failure. This scheme is not used for CPs but is sometimes employed for less critical peripherals. Most of the maintenance and recovery functions of a switch are also controlled from this level.

5.3.7 Digital Switching System Software Classification

A conceptual diagram of typical digital switching system software is shown in Fig. 5.2. The basic software functionality of a digital switching system can be divided into five basic elements, and other functions can be derived from these basic elements:

- Switching software
- Maintenance software
- Office data
- Translation data
- Feature software

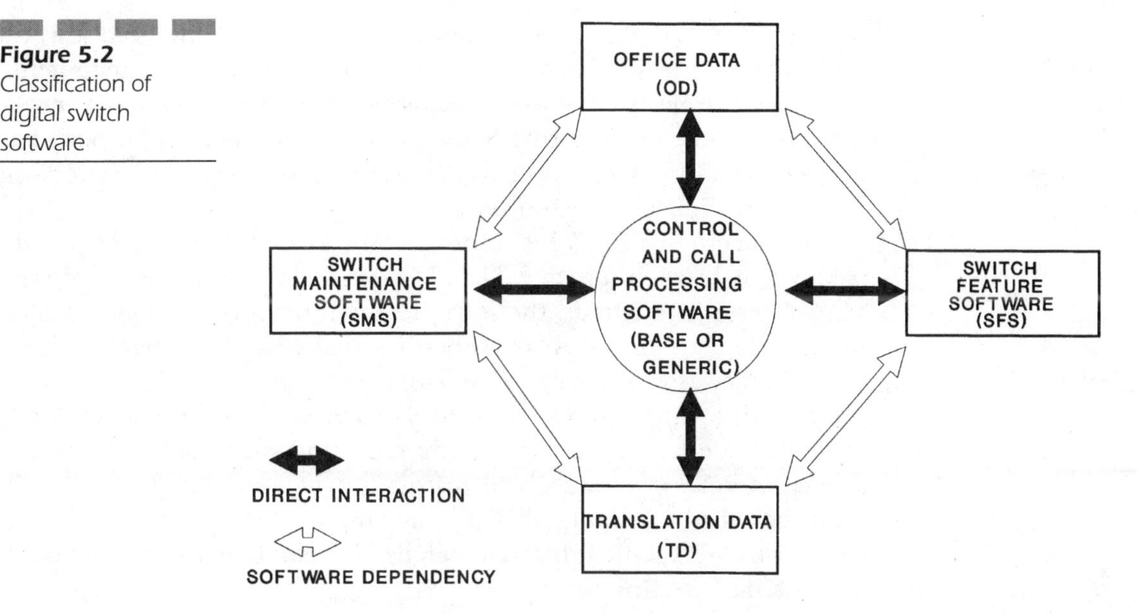

Figure 5.2 Classification of digital switch software

Switching Software The most important layer of software for a digital switching system usually comprises

- Call processing software
- Switching fabric control software
- Network control software
- Periphery control software

Switch Maintenance Software This set of programs is used to maintain digital switch software and hardware. Examples of these types of programs include digital switch diagnostics, automatic line tests, system recovery, patching, and trunk tests.

The recovery software of a modern digital switching system is usually distributed among its subsystems, since most digital switches have a quasi-distributed architecture. This strategy allows the system to recover more efficiently. In earlier SPC systems, the recovery scheme required the entire switching system to go down before it could be reinitialized to a working configuration. The recovery process of a hypothetical digital switching system is covered in Chap. 9.

A digital switching system may employ a large number of programs that are external to the operation of the digital switch, such as operational support systems (OSSs), operator position support, and advanced features (e.g., ISDN, SCP, AIN). These are not shown in Fig. 5.2 as separate items, since they can be external to a digital switching system or may be implemented as a supported feature. Some parts of OSSs can even be viewed as part of digital switching system maintenance software. The networking application of digital switching systems is covered in Chap. 8.

The objective of Fig. 5.2 is to provide the analyst with a clear picture of the digital switch software. The objective of this chapter is to help the analyst better understand the software environment of a digital switch without getting distracted by functions that may not directly impact the reliability assessment of a class 5 digital switch.

The importance of software tools such as compilers, assemblers, computer-aided software engineering tools, and methodologies that are needed to develop, produce, and maintain digital switching system software should not be ignored. They can impact the quality of software. Chapter 6 covers some important details of software reliability and quality assessment techniques.

Office Data The generic program, as described earlier, requires information that is specific to a particular digital switch to operate properly. Digital switching systems have suffered outages due to wrong or improperly defined office data. The easiest way to visualize office data is by comparing them to your personal computer (PC). For the PC to operate properly, the OS has to know what type of color monitor the PC is equipped with, so that correct drivers are installed; the size and type of hard disk installed so that it can access it correctly; types of floppy disks; mouse; and CD ROM. Similarly, the office data of a digital switching system describe the extent of a central office (CO) to the generic program. However, the office data are much more involved and also define software parameters along with hardware equipment.

Some common hardware parameters are

- Number of NCP pairs in the CO
- Number of line controllers in the CO
- Maximum number of lines for which the CO is engineered
- Total number of line equipment in the CO
- Maximum number of trunks and types of trunks for which the CO is engineered
- Total number of trunks of each type in the CO

- Total number and types of service circuits in the CO, such as ringing units, multifrequency (MF) receivers and transmitters, and dial-pulse (DP) receivers and transmitters

These are some examples of software parameters:

- Size of automatic message accounting (AMA) registers
- Number of AMA registers
- Number and types of traffic registers
- Size of buffers for various telephony functions
- Names and types of features supported

These types of parameters are digital switching system-specific and CO-specific. The parameters can literally number in the hundreds and are generated from engineering specifications of a CO.

Translation Data The translation data, also referred to as *subscriber data,* are subscriber-specific and are required for each subscriber. This type of data is generally generated by the telephone companies and not by the suppliers. In some cases, the suppliers may input translation data supplied by the telephone companies. However, the database and entry system for the translation data is supplied as part of the digital switching system software. Typical translation data may consist of:

- Assignment of directory number to a line equipment number
- Features subscribed to by a particular customer, such as call waiting, three-way calling, and call forwarding, etc.
- Restrictions for a particular customer, such as incoming calls only, no long-distance calls, certain calls blocked
- Three-digit translators that route the call based on the first three digits dialed
- Area-code translators that translate the call to a tandem office for 1+ call, which is followed by 10 digits
- International call translators that route the call to international gateway offices based on the country code dialed

Again, literally hundreds of translation tables are built for a CO before it can become functional. If the CO is a new installation, much of the information is provided by the traffic department of a telephone company. The data tables are generated in conjunction with the specification of a new CO. However, if the CO is a replacement for an earlier CO, then all existing data may be required to be regenerated in a different format for the new CO.

Feature Software As mentioned earlier, most features implemented in modern digital switching systems are offered through feature packages. Some of the feature packages are put in a feature group and are offered in a certain market or to a group of telephone companies. These features may be included in the base package of a generic release or, offered as an optional package. In either case, most of the features are considered to be applications for a digital switch. They are engineered to be modular and can be added to a digital switch according to the requirements of the telephone company and associated CO. Some examples of feature packages are

- Operator services
- Centrex feature
- ISDN basic rate
- STP extensions
- SCP database

Depending on the digital switching system, these feature packages can be extensive and large. The analyst of digital switch software should assess the extent of the feature package and its compliance with the requirements of telephone companies.

Software Dependencies Most telephony features of digital switching systems require specific office data and translation data for their operation. They depend on the generation of feature-specific office data and/or translation data. These dependencies are, of course, design-specific. Similarly, the maintenance programs may require a set of specialized office data and/or translation data for testing various functionalities of a digital switching system. These relationships are shown as a software dependency in Fig. 5.2, and direct interactions of a generic program are shown as solid arrows.

5.4 Call Models

The concept of call models is an important one in the design of telephony systems. In its most basic form, the call model describes hardware and software actions that are necessary for connecting and disconnecting a call. Once a basic model for a particular type of call is established, features can then be designed to conform to a particular call model. A complex telephony system may invoke different call models for various types of calls.

Some elements of a basic call model is shown in Fig. 5.3. This model shows two basic states: connect and disconnect. This model by no means represents a complete call model for a digital switching system but is presented here to provide some basic concepts of call models and associated features.

5.4.1 Connect Sequence

The connect sequence consists of software routines that scan the line and detect request for originations. Once the line equipment informs the line scanning program that a line has gone off-hook and that this is a legitimate request for a dial tone, not a hit on the line, the off-hook detection program passes on the control to the test line program. The function of the test line program is to test for the presence of false-ground, high-voltage, line cross, and other conditions that could be detrimental to

Figure 5.3
A basic call model

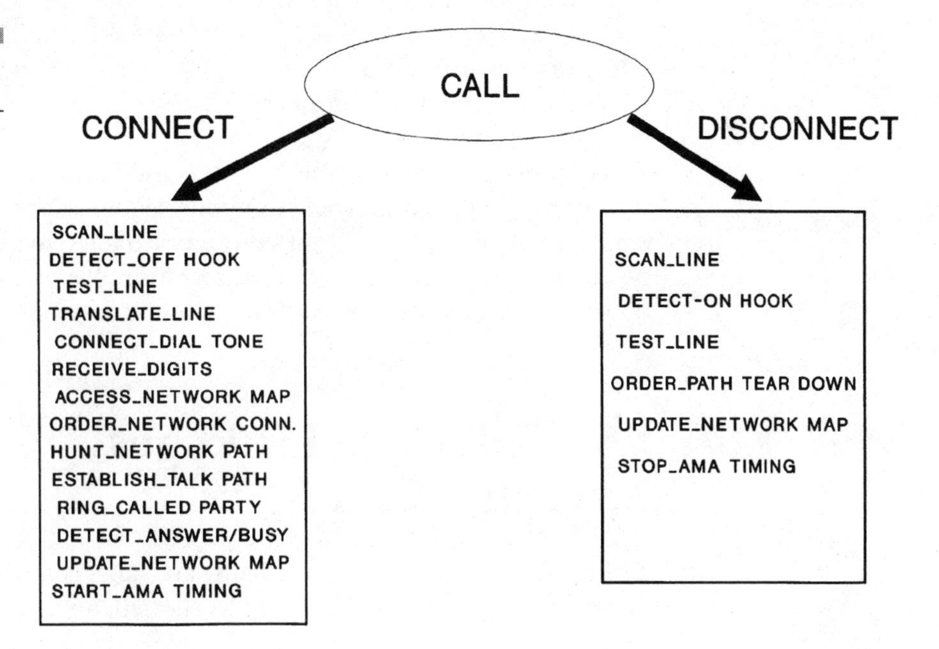

the switching equipment. After successful completion of these tests, a dial tone is returned to the subscriber, signaling the customer to commence dialing. According to customer requirements, all this action needs to be completed in less than 3 seconds. The term *slow dial tone* is used if a switch takes more than 3 seconds to provide the dial tone due to heavy traffic conditions or hardware or software problems. Once the switch detects the start of dialing, the dial tone is removed, and the digit receiver is attached to the line equipment to receive the dialed digits. After the correct number of digits is received by the digit receiver, the switching fabric map is consulted, which tracks the status of all calls and available paths through the switching fabric. Network connect orders are then issued to establish a talking path through the switching fabric. After the completion of the path, the ringing service circuit is connected to the called party, and ringing is initiated. When the called party answers the call, the switching network map is updated and the automatic message accounting timing for billing the call is started.

5.4.2 Disconnect Sequence

The disconnect sequence is also shown in Fig. 5.3. The lines are constantly scanned for connects and disconnects. Once an on-hook or a disconnect is detected, the line is again tested for hazards and a disconnect switching network order is issued to "tear down" the call. Once this is accomplished, the switching network map is updated and the AMA or billing timer is stopped.

As mentioned earlier, the connect and disconnect scenarios mentioned are only two of the numerous sequence models that could be employed in processing a call through a digital switching system. In a typical digital switching system, a large number of call types exist, and they are all associated with various call models.

5.5 Software Linkages during a Call

Earlier chapters introduced the hardware subsystems of a digital switching system. The software linkages to these hardware subsystems will be discussed now. An example of possible software linkages required during a typical call is shown in Fig. 5.4. The line control programs scan the status of lines via the line modules and report the status to the network status program, which in turn works with the network control programs. The line control program also works with the line service circuit programs in providing dial tone, digit receivers, ringing circuits, etc., to the subscriber lines. The network control program orders a network connection through the switching fabric when a subscriber goes off-hook and completes the dialing of all digits for a call. The call processing programs are usually responsible for call processing functions and interface with the feature programs, translation and office data, and automatic message accounting and maintenance programs. The maintenance program is responsible for system recovery, system diagnostics, backup, and other maintenance-related functions. All these functions are available during call processing. Once the call processing program determines for the subscriber line the allowed features and attributes, it allows a call to be established through the switching fabric. The called subscriber may reside in the calling subscriber's digital switch or may be in another digital switch. If the called subscriber is not in the same digital switch, then an outgoing trunk is used to establish a connection to the other digital switch or tandem office. Under this condition, the proper type of outgoing trunk

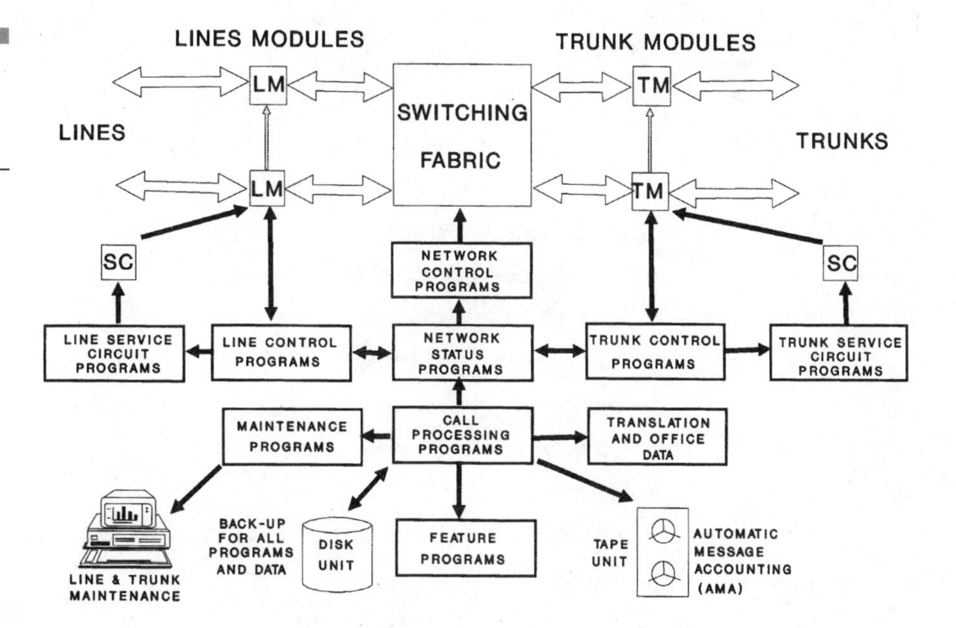

Figure 5.4
Software linkages required during a typical call

is selected and assigned a proper trunk circuit for signaling and supervision. When the called subscriber answers the phone, a talking path is established through the switching network while the line and the trunks are constantly scanned for disconnect from either side. If the subscriber resides in the same digital switch, the special internal line and trunk circuits are used to complete and monitor the call. See Fig. 5.3 for a basic call model. If the called subscriber resides in the same digital switch, the call is classified as an *intraoffice* call; if the called subscriber resides outside the digital switch, the call is termed an *interoffice* call.

5.6 Call Features

The basic function of an end office digital switching system is to provide telephony services to its customers. In today's competitive environment, a feature-rich digital switching system has a competitive edge. Most of the basic features used in the North American market are defined in Bellcore's LSSGR requirements.[5] These requirements are currently divided into the following categories:

- Residence and business customer features
- Private facility access and services
- Attendant features
- Customer switching system features
- Customer interfaces
- Coin and charge-a-call features
- Public safety features
- Miscellaneous local system features
- Interoffice features
- Call processing features
- Database services
- Data services
- System maintenance features
- Trunk, line, and special service circuit test features
- Administrative features
- Cut-over and growth features
- Billing and comptrollers features

Clearly, just from the categories listed above, a digital switching system usually supports a large number of features. Each category may contain scores of features. A modern digital switching system may contain several hundred features.

5.6.1 Feature Flow Diagrams

The features employed in a digital switching system are usually very complex, and flow diagrams can help one to understand their functionalities. A simplified flow diagram for one of the most commonly used subscriber features, call forwarding (CF), is shown in Fig. 5.5. This feature has three modes of operation:

Feature Activation The feature is activated when the customer goes off-hook and dials an activation code. The software checks for the correct validation code. If the activation code is wrong, the subscriber does not get the second dial tone. If the activation code is correct, the subscriber gets a second dial tone and is allowed to dial the call-forwarding telephone number. The call-forwarded subscriber line is rung once, and the number is recorded in the system memory for future use.

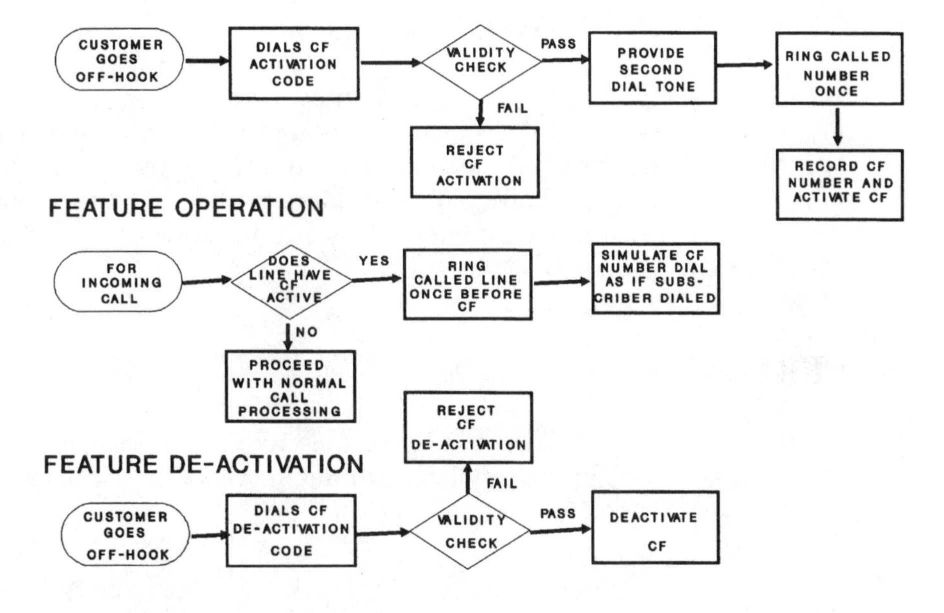

Figure 5.5 Simplified flow diagram for call-forwarding (CF) feature

Feature Operation Now, suppose the subscriber receives a call on the line that has the CF feature activated. The system rings the called subscriber once and then forwards the call to a number previously recorded by the subscriber during feature activation.

Feature Deactivation This feature can be deactivated when the subscriber goes off-hook and dials the deactivation code. If the code is valid, the CF number is removed; otherwise, the deactivation request is ignored.

Note that this was a very simplified flow diagram for a feature. The actual flow diagrams for some of the features are far more complex.

5.6.2 Feature Interaction

One of the obvious problems that can arise owing to the existence of so many features on a single digital switching system is feature interaction. This can happen even in the most advanced and best-designed digital switches. One way to minimize this problem is to conduct regression tests on the software and related hardware. This subject is discussed in greater detail in Chap. 6, which addresses software reliability and quality assessment issues.

5.7 Summary

In this chapter we introduced a basic software architecture of a digital switching system, classification of the switch software, basic call models, software linkages during a call, and feature flow diagrams. This should give the reader a more thorough understanding of hardware, software, and procedural interactions in a digital switching system.

REFERENCES

1. A. Ralston and E. Reilly, *Encyclopedia of Computer Science and Engineering*, Van Nostrand Reinhold Co., New York, 1983, p. 1053.

2. C. Vick and C. Ramamoorthy, *Handbook of Software Engineering*, Van Nostrand Reinhold Co., New York, 1984, pp. 163, 817.

3. A. Stevens, *C++ Database Development*, MIS Press, New York, 1992, p. 34.

4. J. Martin, *Computer Data-Base Organization*, Prentice-Hall, Englewood Cliffs, NJ, 1977, p. 53.

5. *Bellcore's LATA Switching Systems Generic Requirements*, TR-TSY-000504 (LSSGR Features), a module of TR-TSY-000064, Piscataway, NJ.

CHAPTER 6

Quality Analysis of Switching System Software

6.1 Introduction

This chapter will introduce the reader to some important concepts essential for assessing the quality of digital switching system software. Note that the software development process for the digital switching system is complex and varies greatly for different products. The objective of this chapter is not to make the reader into a certified software assessor, but to develop a simple methodology for verifying the effectiveness of processes that are usually employed in developing digital switching software. References to some standardized methodologies that are currently used by the software industry to assess software quality are provided. These methodologies are applicable to any software development process, including digital switching software.

6.2 Scope

The concept of the software life cycle, including requirements capture, low- and high-level design, software testing methodology, and software deployment, is covered in this chapter. Also the Software Engineering Institute's Capability Maturity Model (CMM) and the International Standard Organization (ISO) 9000-3 standard are discussed. A high-level view of a few other software quality-assessing methodologies is given, too.

6.3 Life Cycle of Switching System Software

To develop software, various tasks associated with the development of a software product are divided into a smaller group of activities. These groupings are usually referred to as *phases*, and the entire process of creating a software product defines the software life cycle. Software engineers use many different models for developing software based on the requirements of the business. The objective of this chapter is not to delve into the intricacies of different models but to describe a typical life-cycle model that is usually employed in developing switching system software. Currently, most switching system software is based on the *waterfall model.*[1] This was one of the earliest models adapted for switching system software development. Due to current demands on the industry for the rapid introduction of telephony features, some fast prototyping methods and a few variants of the waterfall model are being used, such as the V model.[2] Regardless of the model used, the life cycle of switching system software, shown in Fig. 6.1, can be grouped into the following four phases:

- Software development
- Software testing
- Software deployment
- Software maintenance

These phases are shown in a waterfall arrangement, meaning that one phase of activity is usually completed before the activities in other phases start. In reality, many variations of this sequence are currently in use. Some switching system suppliers are beginning to use object-oriented programming (OOP) methodologies.[3] These methodologies employ the *spiral*[4] model in which the software is created in small increments until the

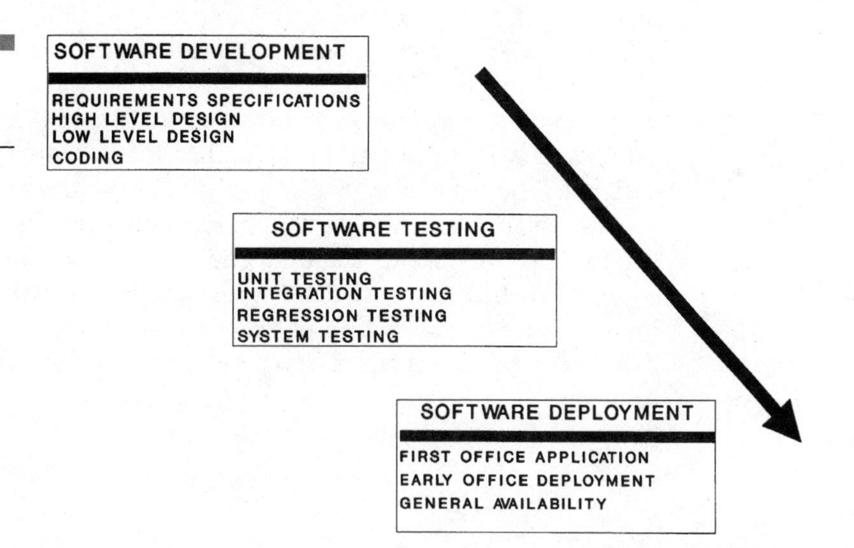

Figure 6.1 Switching system software life cycle

release stage is reached. However, currently these methodologies are not widely used for switching system software, since the maintainability of OOP products in a real-time switching environment is still questionable and requires further research. Therefore, this concept is not covered in this chapter, but references are provided for further study. However, in this author's opinion the future of software development will be highly influenced by OOP methodologies and computer-aided software engineering (CASE) tools.

6.4 Software Development

The front end of the software life cycle is often referred to as *software development.* See Fig. 6.1 for details. It usually starts with a planning stage. In the early days of software development, plans for future software products were conceived by technical management groups. Currently, the marketing department plays a very important role in this area and is usually responsible for understanding future market trends. Once the marketing department and technical groups have identified the market and particular customer needs, a planning document is created. These plans usually describe high-level needs of a software product. Later, more detailed information is captured for the requirements specification stage. This information is further refined during the high-level design. In the next stage of software development, *low-level design,* which is spawned from the high-level design, is enhanced with implementation details. The last stage of the high-end software development life cycle is coding, in which the actual software code is written. Contrary to common perception, the coding phase of software development entails only 20 percent of the entire software life-cycle effort, and much greater resources are required for other phases of the software development process than generally believed.

6.5 A Methodology for Assessing Quality of Switching Software

In the next few subsections, an embryonic questionnaire-based methodology for assessing digital switching system software is outlined. This methodology requires a much deeper understanding of software development processes than can be communicated in this book. The model of this chapter is not intended to replace any existing assessment methodologies. It is simply a methodology for verifying the effectiveness of various software life-cycle phases that are most applicable to digital switching software.

This method is based on asking a set of questions, verifying the responses, and applying a weighting scheme to the responses. Based on the responses, a second set of more direct questions is asked, and the responses are verified and weighted. The key to this method is to have a good understanding of the software architecture of the digital switch under assessment and then to analyze the associated software development process. This book will assist the analyst in accomplishing both. The weighting scheme is covered in the next section.

The next few subsections provide a very high-level view of the software life-cycle stages currently employed for developing digital switching system software, which is followed by a few sample questions that are to be asked during the software assessment process. To conduct a thorough assessment, a large number of additional questions must be developed specific to each software life cycle and product. The questions given here are just a "starter set" to give a feel for the questions that need to be asked, and they do not cover any specific assessment process.

6.5.1 Result Validation and Recording

The results of the validation can be recorded in a simple table, shown in Fig. 6.2. If the answer to the question is yes for 100 percent of projects involved in developing the product, then a score of 1.0 is placed in the first column. Similarly, for 50 percent, a score of 0.5 is placed in the second column and for 25 percent a score of 0.25. When the answer is no, a 0 is placed in the No column. When the question is not applicable, N/A is placed in the N/A column. The N/A column is useful in cases in which a generic set of questions is developed for all products and some questions may not apply in all cases. These questions can then be ignored for specific products. A weight of 1 to 5 is placed in the Weight column based on the importance of the question. The total score is then calculated by multiplying the numbers in the yes and no columns by the appropriate weight and entering the results in the Weighted Score column. The phase total can then be calculated by adding all the

Figure 6.2 Requirements specifications questionnaire

REQUIREMENTS SPECIFICATIONS

VALIDATED ANSWERS → / SAMPLE ASSESSMENT QUESTIONS ↓	YES			NO	N/A	WEIGHT (1 to 5)	WEIGHTED SCORE
	100% OF PROJECTS (1.0)	50% OF PROJECTS (0.50)	25% OF PROJECTS (0.25)	0% OF PROJECTS (0.0)	NOT APPLICABLE (NA)	MULTIPLIER (X)	(WS)
1. *Are any formal methods used for analyzing the requirements once written, like data flow diagrams, data dictionary, functional descriptions, etc. [see ref. 1]*							
2. *Are any prototyping tools used for verifying the requirements with customer participation ?*							
3. *Are customers requirements used as a basis for software product requirements ?*							
4. *Are any requirements capture tools or special languages used in capturing requirements ?*							
5. *Can the software product requirements be traced to the customer's requirements ?*							
• • •							

TOTAL SCORE = $\sum$ (WS)

weighted scores, which in turn can be transformed to a normalized score for each phase. Details of the normalization process are covered in Sec. 6.6, and an overall scoring scheme is presented in Table 6.1.

Note that this is just a suggested way of verifying the effectiveness of a software life-cycle process. Software assessment is a very complex process, and the technique described here by no means represents an alternative to any standard software assessment. It provides an insight into the complexities that need to be understood.

6.5.2 Requirements Specifications

The next stage in developing software after the planning stage is the requirements stage,[5] in which specifications related to a software product are developed. At this stage, customer requirements and internal requirements are written in a manner that can be understood by the software designers. The basic objective for this phase of software development is to capture accurate customer and internal requirements.

One way to understand the supplier's software development methodology is to thoroughly review the supplier's software life-cycle documents. These documents describe the software life cycle and provide details of the software development process. Once this software development methodology becomes apparent, a questionnaire can be created to verify the effectiveness of the software development process. For instance, the very small sample of questions provided below could be used as an initial set of questions for verifying the effectiveness of the requirements capture stage:

1. Are any formal methods used to analyze the requirements once written, such as data flow diagrams, a data dictionary, or functional descriptions?[6]
2. Are any prototyping tools used to verify the requirements with customer participation?
3. Are customer requirements used as a basis for software product requirements?
4. Are any requirement capture tools or special languages used in the capturing requirements?
5. Can the software product requirements be traced to the customer's requirements?

Depending on the complexity of the software development system and the depth and breath of assessment, this set of questions can be

Life Cycle Phase	Number of Questions N Phase	Number of N/A Questions	Effective Number of Questions N — NA = (E)	Maximum Possible Score E • 5 = M	Total Score Score Weighted Σ (WS)	Normalized Score $\frac{\Sigma\ WS \cdot 5}{M}$
Sofware development						
Software testing						
Software deployment						
Normalized overall score for all life-cycle phases						Σ(Normalized score) ÷ (Total number of life-cycle phases)

TABLE 6.1

A normalized scoring table

very large indeed. Once these questions are asked, the answers must be validated by observing the process itself while assessing the effectiveness of a software development process. The questions are then presented in an assessment table similar to the one shown in Fig. 6.2 to ascertain the effectiveness of the requirements specification process.

6.5.3 High-Level and Low-Level Design

After the capture of software requirements, a process is needed to translate requirements to software design.[6] This process is generally referred to as *high-level design* (*HLD*). The basic objective of this level of design is to define the software architecture and data elements that are necessary to satisfy the specifications described in the requirements stage of the software life cycle. Once the HLD is completed, it is usually followed by a refinement process in which HLD is expanded to cover greater details, e.g., defining data structures, mathematical equations, algorithms, and other details needed for writing the software code. This process is generally referred to as *low-level design* (*LDL*). Again, as mentioned earlier, it is not the objective of this chapter to make the reader an expert in software design analysis, which in itself is a vast and complex subject. References are provided at the end of this chapter for in-depth reading. A list of sample questions is given here to illustrate the types of questions that need to be asked. These questions are not all-inclusive or exhaustive for a thorough analysis of the design process.

1. Is the design process formally documented, and does it use well-defined design methodologies?
2. Does the supplier conduct design reviews on a regular basis?
3. Does the design methodology separate data and procedure in a consistent manner?
4. Does the design methodology clearly define module functionalities and module dependencies?
5. Are designs written in a manner that would allow traceability between requirements, design, and code?

A much more detailed list of questions needs to be developed than is shown above, to fully assess the design process. The answers can then be recorded in an assessment table similar to the one shown in Fig. 6.3 to ascertain the effectiveness of the design process.

Figure 6.3
High/low-level design questionnaire

HIGH/LOW LEVEL DESIGN

VALIDATED ANSWERS → / SAMPLE ASSESSMENT QUESTIONS ↓	YES			NO	N/A	WEIGHT (1 to 5)	WEIGHTED SCORE
	100% OF PROJECTS (1.0)	50% OF PROJECTS (0.50)	25% OF PROJECTS (0.25)	0% OF PROJECTS (0.0)	NOT APPLICABLE (NA)	MULTIPLIER (X)	(WS)
1. Is the design process formally documented and use well defined design methodologies ?							
2. Does the supplier conduct design reviews on a regular basis ?							
3. Does the design methodology separate data and procedure in a consistent manner ?							
4. Does the design methodology clearly define module functionalities and module dependencies ?							
5. Are designs written in a manner that would allow traceability between requirements, design and code ?							
• • • •							

TOTAL SCORE = $\sum$ (WS)

6.5.4 Coding

The next step in developing software is the coding. Most people refer to this phase as *program creation* or *programming*. During this stage of the software life cycle, low-level design is converted to actual software codes that are implementable on computing elements such as processors and microprocessors. Software codes for digital switching systems are usually written in high-level programming languages such as C, C++, PASCAL, and CHILL. Programs written in higher-level languages referred to as *source code* are then compiled to an intermediary object code which in turn is translated to machine code(s). Machine codes are processor-specific implementations of a programming instruction. This implies that given a source code, object codes for different target processors can be produced with appropriate compilers. Again the object of this subsection is to point out some areas in the coding phase[7] of software development that need to be understood and considered from a software assessment angle.

1. Does the supplier have documented coding standards and conventions?
2. Does the supplier follow these standards and conventions consistently?

3. Does the supplier conduct code reviews frequently?
4. Does the supplier place all the source codes under a configuration[8] management system?
5. Does the supplier employ certified compilers for producing object codes?

A much more detailed list of questions needs to be developed to fully assess the design process. The answers can then be entered in an assessment table similar to the one shown in Fig. 6.4 for ascertaining the effectiveness of the coding process.

6.5.5 Software Testing

Software testing is a process that removes defects from software before its release to customers. One can also view software testing as a process that validates the requirements upon which the development of the software was based. Software testing can also be viewed as a process that validates software behavior with realistic data.

A Typical Digital Switch Software Test Methodology A typical digital switch software test methodology is shown in Fig. 6.5a. A test

CODING

VALIDATED ANSWERS → / ↓ SAMPLE ASSESSMENT QUESTIONS	YES			NO	N/A	WEIGHT (1 to 5)	WEIGHTED SCORE
	100% OF PROJECTS (1.0)	50% OF PROJECTS (0.50)	25% OF PROJECTS (0.25)	0% OF PROJECTS (0.0)	NOT APPLICABLE (NA)	MULTIPLIER (X)	(WS)
1. Does the supplier has documented coding standards and conventions ?							
2. Does the supplier follow these standards and conventions consistently ?							
3. Does the supplier conduct code reviews frequently ?							
4. Does the supplier place all the source codes under a configuration management system ?							
5. Does the supplier employ compilers for producing object codes ?							
• • • • •							

TOTAL SCORE = $\sum$(WS)

Figure 6.4 Coding questionnaire

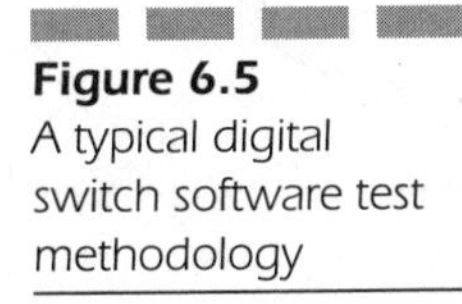

Figure 6.5
A typical digital switch software test methodology

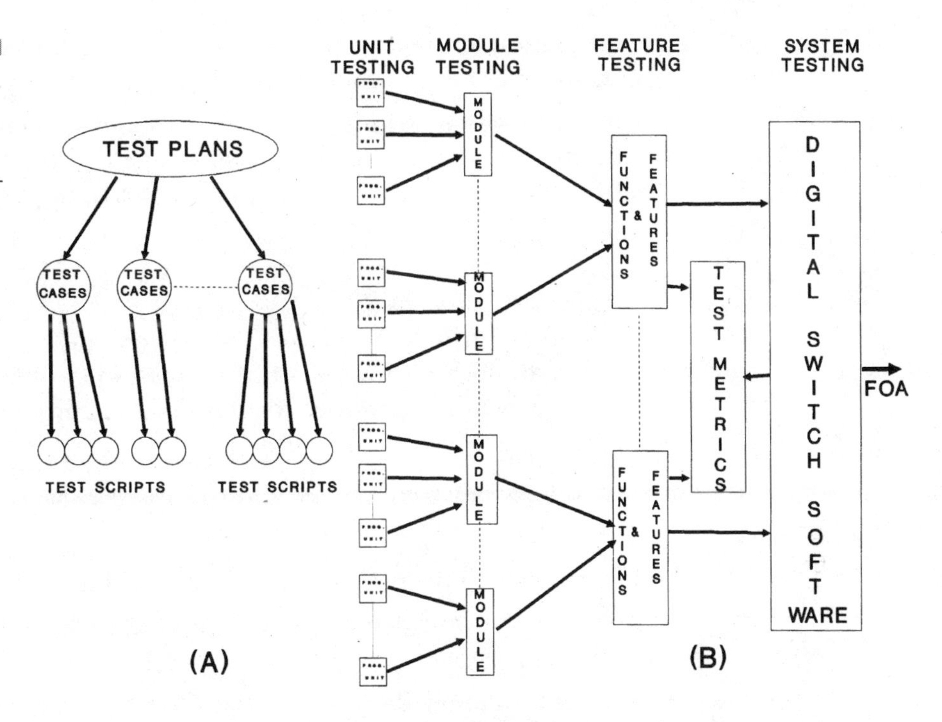

process is conceived via test plans. Test plans represent high-level plans and strategies for testing a software product. Test cases contain test information with more granularity than test plans; for instance, a test plan may set strategies for testing all outgoing trunks while a test case will define exact testing parameters for an outgoing operator trunk. The test scripts represent the exact executable tests, usually written in a test scripting language which can be automatically or manually executed on a test bed to which the product under test is attached.

Most suppliers of digital switching systems employ testing techniques that are best suited to their products. However, most of the current techniques can be mapped into the following four test life cycles (Fig. 6.5 *b*):

- *Unit testing:* The first very basic set of tests is usually conducted on newly developed software modules that contain a set of programs or program units, sometimes referred to as sections, subprograms, etc. These tests are usually conducted by software developers.
- *Integration testing:* The next-higher level of testing checks the interworking of software modules. These tests are usually conducted by groups other than software developers, but the software developers act as liaisons during the process. Most of

integration testing is usually conducted in a laboratory environment with simulated hardware and software.

- *Regression testing:* Regression testing is usually conducted throughout the testing life cycle to ensure that the features that functioned correctly in the last release continue to function correctly in the new release.
- *System testing:* This is the last phase of testing before the software is released to the customer, also referred to as *alpha* testing in the industry. During this phase of testing, the software is exposed to real hardware and simulated traffic conditions. The stability of software is also determined to assess its suitability for release.

These sample questions merely illustrate the types of questions that need to be asked; they are not all-inclusive or exhaustive for a thorough analysis of the software testing process.

1. Is the testing life cycle well documented and formalized?
2. Are the documented procedures adhered to during the software testing process?
3. Does the supplier write formalized test plans, test cases, and test scripts for conducting software tests?
4. Does the supplier conduct regression testing on features that are to be re-released with new generic programs?
5. Is the system test organization an independent organization with powers to accept or reject a release based on testing results?

A much more detailed list of questions must be drafted than that shown above to fully assess the testing process.[9] The answers can then be entered in an assessment table similar to the one shown in Fig. 6.6 for ascertaining the effectiveness of the testing process.

6.5.6 Software Deployment

The software deployment process for a digital switching system is a very important aspect of the software life-cycle phase. It can impact the operational reliability of a working digital switch during the deployment process as well as its future operational capability. Before a release is made generally available for all central offices (COs) in the field, the following three modes of release are usually considered. Different suppliers may call them by different names, but they all involve prerelease testing of soft-

Figure 6.6
Testing questionnaire

COMMON

VALIDATED ANSWERS → / ↓ SAMPLE ASSESSMENT QUESTIONS	YES			NO	N/A	WEIGHT (1 to 5)	WEIGHTED SCORE
	100% OF PROJECTS (1.0)	50% OF PROJECTS (0.50)	25% OF PROJECTS (0.25)	0% OF PROJECTS (0.0)	NOT APPLICABLE (NA)	MULTIPLIER (X)	(WS)
1. Is the testing life cycle well documented and formalized ?							
2. Are the documented procedures adhered to during the software testing process ?							
3. Does the supplier write formalized test plans, test cases, test scripts for conducting software tests ?							
4. Does the supplier conduct regression tests on features that are to be re-released with new generic program ?							
5. Is the system test organization an independent organization with powers to accept or reject a release based on testing results ?							
• • • • • •							

TOTAL SCORE = $\sum$ (WS)

ware at the customers' locations. The objective of deployment testing is to ensure that the release functions satisfactorily in an operational environment for a period of time before it is released for general use. Currently, the following strategies are used for deploying switching system software:

First Office Application (FOA) This can be considered an extension of system testing. This is the very first exposure of a new release to an in-service digital switching system with live telephony traffic. Some suppliers consider this to be beta testing and the CO where the FOA is carried as a beta or FOA test site. This phase of deployment is considered essential since not all elements of a release can be adequately tested during system testing with all possible combinations of deployed hardware and software features and with complex CO traffic patterns that cannot usually be simulated in a system test laboratory. A release is put through the FOA process, in which an adequate amount of time is allowed to pass to see the impact of fault corrections or patches implemented during FOA period. After this "absorbing" period, the release is made available for limited distribution to other sites for further testing with specific hardware and feature combinations.

Early Office Application (EOA) This phase of deployment can be considered an extension of FOA, and some suppliers may institute multiple

FOAs instead of EOA. The basic object of EOA is to validate a release with different combinations of hardware and features that could not generally be tested during FOA. This is possible since some hardware (vintage or new) may not have existed in the FOA office or some specific features may not have been ordered or required for a particular FOA site. Usually about half a dozen sites may be required for the EOAs depending upon the size and new-feature content of a release.

General Availability (GA) This is the final phase of deployment. It is assumed that all the faults found during FOA and EOA are corrected and implemented in the release and that the release has no known critical or major faults. Usually some time is allowed to elapse between FOA, EOA, and GA for the release to stabilize, referred to as the *soak period,* before the release is made generally available to all sites. It is assumed that when a GA release is installed, all customer documentation, customer support, and future installation time lines have been established and that the release is stable.

A list of sample questions is provided here:

1. Is the software deployment process documented and formalized?
2. Does the supplier select the FOA central office well in advance that meets FOA requirements and objectives?
3. Does the supplier support a methodology that allows the concept of joint testing with customers?
4. Does the supplier support a formalized fault recording and fault correction methodology that is used during the FOA and EOA time frame?
5. Does the supplier support a formalized acceptance test methodology that a customer can use to assess proper operation of a release before acceptance?

An assessment table similar to the one shown in Fig. 6.7 can be used to ascertain the effectiveness of the deployment process. References are provided at the end of this chapter for in-depth reading.

6.6 Overall Scoring

A simple overall scoring scheme, shown in Table 6.1, can be utilized once the assessment of all software life-cycle phases is completed. As

Figure 6.7
Deployment questionnaire

COMMON

VALIDATED ANSWERS → / ↓ SAMPLE ASSESSMENT QUESTIONS	YES			NO	N/A	WEIGHT (1 to 5)	WEIGHTED SCORE (WS)
	100% OF PROJECTS (1.0)	50% OF PROJECTS (0.50)	25% OF PROJECTS (0.25)	0% OF PROJECTS (0.0)	NOT APPLICABLE (NA)	MULTIPLIER (X)	
1. Is the software deployment process documented and formalized?							
2. Does the supplier select the FOA Central Office well in advance that meets FOA requirements and objectives?							
3. Does the supplier support a methodology that the concept of joint testing with customers?							
4. Does the supplier have a formalized fault recording and fault correction methodology that is used during the FOA/EOA time frame?							
5. Does the supplier support a formalized acceptance test methodology that a customer can use to assess proper operation of a release before acceptance?							
• •							

TOTAL SCORE = $\sum$ *(WS)*

shown in the table, all the results can be tabulated and normalized, and the final score will be between 1 and 5, with a higher number representing a better software process. Naturally, the final judgment of the scores will be entirely based on the assessor's view of the process and product; this is just a simple methodology designed to aid judgment.

6.7 Two Important Software Assessment Models

Currently, two software assessment models are in general use, one developed by the Software Engineering Institute (SEI) at Carnegie Mellon University and called the Capability Maturity Model (CMM), and the other by the International Standards Organization Technical Committee ISO/TC 176. Both models support a certification process. The object of the next two subsections is merely to introduce these models; it is not intended to be an exhaustive treatment of the subject.

6.7.1 The CMM

The CMM[10] uses a continuous improvement process as the basis for improving the maturity of the software development process. It divides maturity into five distinct levels, identified as CMM level 1 to CMM level 5. The first level is referred to as the *initial* or ad hoc level, the second level as *repeatable*, the third level as *defined*, the fourth level as *managed*, and the last or fifth level as *optimizing*. For a full understanding of CMM levels, consult Ref. 10. This is a high-level view of the five CMM levels.[11]

- *Level 1* At this level the software development and maintenance environment is not stable. Sound management practices are not used. The process can be described as chaotic, based on the competence of individuals and not that of an organization. This type of organization develops software which is often off-schedule and off-budget.
- *Level 2* At this level software policies and procedures are established. Software requirements are baselined, and the organization takes a disciplined approach to software development. Planning and tracking of projects are stable, allowing the successful practices of one project to be repeated in other projects.
- *Level 3* At this level of software maturity, the software process is well defined. A well-defined process includes standards and procedures for performing work, input/output criteria, peer reviews, and completion criteria. A level 3 organization has stable and repeatable processes for its engineering and management activities. The entire organization has a common understanding of defined software processes.
- *Level 4* In the managed level, the processes are instrumented or measured. Quality goals are set for the product as well as the process, and variations are controlled. At this maturity level, the quality of process and of product is predictable within certain bounds. The process at this level is stable and measured, and actions can be taken to correct process or quality problems.
- *Level 5* At the optimizing level, the entire organization uses a continuous-improvement program (CIP). The organization also uses defect analysis and prevention techniques that reduce rework. A level 5 organization continuously strives to improve its process capability, which consequently improves project performance. The

organization also uses incremental advancements to existing processes with new methods and techniques.

An organization's performance can be ascertained against detailed requirements of these levels by an SEI-certified team, and its software can be assigned a maturity level. The objective of the organization is to keep improving its CMM level.

It is an effective method for assessing the software maturity of a software development process and is gaining in popularity. However, CMM levels do not cover particular application domains such as digital switching software development or advocate the advantages of specific technologies such as object-oriented programming or the use of computer-aided software engineering tools.

6.7.2 The ISO Model

The ISO 9000-3 is a European standard[12] and is a guideline for quality management and quality assurance for software development, supply, and maintenance. This guideline helps in verifying the effectiveness of the software quality system but not the software itself. This guideline describes precise requirements related to

- The quality system framework such as management responsibility, quality policy, joint reviews, and corrective actions
- Life-cycle activities such as design, testing, reviews, and installation
- Supporting activities such as configuration management, documentation, measurement, and training

On the basis of these requirements, a software organization is "audited" by an ISO-certified organization which is authorized to audit and issue a certificate of ISO compliance. To maintain the ISO certification, an organization has to be recertified at certain intervals.

The advantage of the ISO 9000 standard is that it contains a set of high level requirements that an organization must meet and is useful as a baseline for a supplier's software quality system. An ISO-certified software organization undoubtedly will have a well-documented software life-cycle process. The disadvantage is that customers assume that an ISO-certified supplier produces high-quality software, which may not be true. Although an ISO-certified supplier has a certified software quality system, this does not guarantee great software since good software depends on many factors, such as the use of requirement capture tools, innovative design techniques, CASE tools, new testing strategies, etc., which are not covered in depth by the ISO standard.

6.7.3 Bellcore's RQGR Methodology

Assessment of software quality can be made by using the two models discussed above or other standards, such as Bellcore's *Reliability and Quality Generic Requirements,*[13] or the Department of Defense's *Defense System Software Quality Program.* Bellcore's generic requirements for software quality assessment were developed for the Bell Operating Companies and assesses the quality system

- Framework—covers management responsibility, system audits, corrective audits, corrective actions, etc.
- Life-cycle activities—cover requirement specifications, development and quality planning, testing and validation, maintenance, etc.
- Supporting activities—cover configuration control, document control, quality records, measurements, tools and techniques, etc.

6.8 Summary

This chapter introduced some basic concepts of software development processes that are commonly used to develop digital switching system software. The concept of software life-cycle development phases along with software testing was covered in detail. A questionnaire-based methodology for assessing software quality was introduced. A synopsis of other software assessment methodologies such as CMM and ISO 9000 was presented.

REFERENCES

1. C. R. Vick and C. V. Ramamoorthy, *Handbook of Software Engineering,* Van Nostrand Reinhold, New York, 1984.

2. William W. Agresti, *New Paradigms for Software Development,* IEEE Computer Society, 1986.
3. James Rumbaugh, Michael Blaha, William Premerlani, Frederick Eddy, and William Lorensen, *Object Oriented Modeling and Design,* Prentice-Hall, Englewood Cliffs, NJ, 1991.
4. Ivar Jacobson, *Object-Oriented Software Engineering, A Use Case Driven Approach,* Addison-Wesley, Reading, MA, 1992.
5. Donald C. Gause and Gerald M. Weinberg, *Exploring Requirements—Quality before Design,* Dorset House Publishing, 1989.
6. R. S. Pressman, *Software Engineering, A Practitioner Approach,* McGraw-Hill, New York, 1987, p. 164.
7. R. Dunn and R. Ullman, *Quality Assurance for Computer Software,* McGraw-Hill, New York, 1982.
8. Steve J. Ayer and Frank S. Patrinostro, *Software Configuration Management,* McGraw-Hill, New York, 1992.
9. William E. Perry, *A Structured Approach to Systems Testing,* QED Information Sciences, 1988.
10. Mark C. Paulk, Bill Curtis, Mary Beth Chrissis, and Charles V. Weber, *Capability Maturity Model for Software,* CMU/SEI-93-TR24, Software Engineering Institute, 1993.
11. Carnegie Mellon University, Software Engineering Institute, *The Capability Maturity Model, Guidelines for Improving the Software Process,* Addison-Wesley, Reading, MA, 1995.
12. TickIT, *Guide to Software Quality Management System Construction and Certification Using EN29001,* Department of Trade and Industry, London, U.K.
13. *Bellcore's Reliability and Quality Generic Requirements,* TR-NWT-000179, issue 2, June 1993, Piscataway, N.J.

FURTHER READING

Bellcore's In-Process Quality Metrics (*IPQM*), GR-1315-CORE, issue 1, September 1995.

Bellcore's Reliability and Quality Measurements for Telecommunications System (*RQMS*), GR-929-CORE, issue 1, December 1995.

Defense Department Software Development, DOD-STD-2167A, Department of the Navy, Washington, April 1987.

Perry W.: *A Structured Approach to Systems Testing,* 2d ed., QED Information Sciences, Inc., Wellesley, MA, 1988.

Standard for Developing Life Cycle Processes, IEEE Standard 1074, Institute of Electrical and Electronics Engineers, New York, NY, 1991.

CHAPTER 7

Maintenance of Digital Switching Systems

7.1 Introduction

After the digital switching system is installed, switch maintainability becomes an important consideration. This chapter introduces some basic information that is needed to assess the maintainability of a central office (CO).

7.2 Scope

This chapter introduces typical interfaces that are utilized in maintaining central offices both remotely and locally. Topics essential to CO maintenance such as fault reports, software patches, and the software and hardware upgrade process, including firmware, are also covered.

7.3 Software Maintenance

Proper software maintenance is again a vast subject and in this author's opinion a neglected one. The software industry spends almost 80 percent of its effort in maintaining software, but not enough research has been conducted to improve software maintainability. From the digital switching point of view, digital switch maintainability can be grouped into two broad categories:

- *Supplier-initiated software maintenance:* This consists of software maintenance actions needed to update or upgrade a generic release of a digital switch. These also include applications of "patches" or software corrections that are required to correct faults in an existing generic release.
- *Software maintenance by site owners:* These are routine maintenance actions that must be performed by the owners of a digital switch to keep it operational. Examples could be routine diagnostics, updating of translation tables, and addition of lines and trunks to a digital switch.

7.4 Interfaces of a Typical Digital Switching System Central Office

Most of the common interfaces needed for a digital switching system central office are shown in Fig. 7.1. The maintainability of a CO depends on satisfying the needs of all these and other interfaces. A group of COs is usually assigned to a switching control center (SCC), in the Bell Operating Companies environment, but local maintenance personnel are also involved in maintaining COs. The next level of maintenance is assigned to the electronic switching system assistance center (ESAC) in parallel with the maintenance engineers. Maintenance engineers are not involved with daily maintenance but oversee resolution of recurrent maintenance issues. The ESAC organization usually controls generic upgrades, patching, operational trouble reports (OTRs), and interfaces with the supplier's regional technical assistance centers (RTACs) and technical assistance centers (TACs) to solve unusual and difficult maintenance problems. Note that this is only a typical arrangement and will vary with telephone companies and switching system products. But most telephone companies support different

Figure 7.1 Organizational interfaces of a typical CO

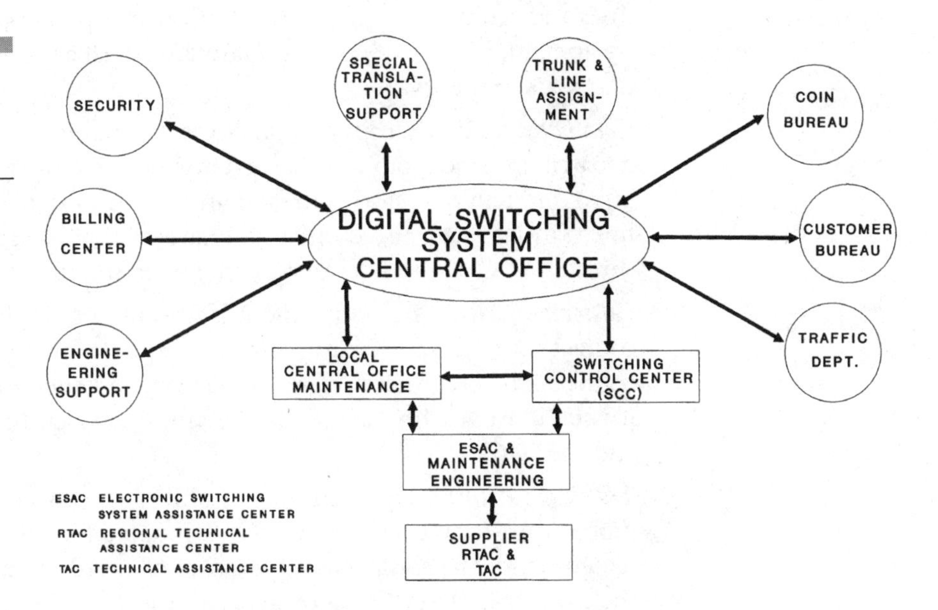

levels of digital switch maintenance. These other departments interact with a digital switch:

- *Engineering support:* This department writes specifications for a new digital switch and engineers' additions to the existing CO. This department also interfaces with the supplier's engineering department, CO plant department, and traffic department with the objective of issuing accurate engineering specifications for a new digital switch installation or addition.
- *Billing center:* The billing center is responsible for processing automatic message accounting (AMA) or billing tapes from a CO to produce customer bills. Currently, billing information can also be transmitted directly to the billing center.
- *Security:* This department provides security services for the digital switching system to prevent unauthorized entry and fraudulent use of the telephone service.
- *Special translation support:* This group provides support in establishing unusual translations for COs that provide special services for large corporations with complete call routings, trunk translations, etc.
- *Trunk and line assignment:* This group's main function is to assign lines and trunks to a digital switch's line equipment and trunk equipment, respectively. It also maintains database of line and trunk assignments.
- *Coin bureau:* Usually, coin equipment is maintained by a separate department since coin telephones employ different instruments and often different operators. Special coin collection signals and special line translators are also employed. However, the department works through SCCs and ESACs to correct any coin-related problems.
- *Customer bureau:* This department is usually the single point of contact for telephone customers with requests for telephone connection, disconnection, reconnection, and telephone problems. It usually works through the trunk and line assignment groups and the SCCs.
- *Traffic department:* The main responsibility of this group is to model and study telephony traffic through a digital switch. It recommends the addition and removal of trunks in a CO, based on the dynamics of traffic patterns. The group also interfaces with the

engineering support group concerning trunk estimates necessary for the installation of a new digital switch.

7.5 System Outage and Its Impact on Digital Switching System Reliability

Digital switch outages represent the most visible measure of switching system reliability and affect maintainability. Various studies[1] have been conducted to better understand the causes of digital switch outages. Traditionally, the causes of outages have been classified into four categories:

- *Software deficiencies.* This includes software "bugs" that cause memory errors or program loops that can be cleared only by major initialization.
- *Hardware failure.* This relates to simplex and/or duplex hardware failures in the system which result in a system outage.
- *Ineffective recovery.* This category includes failure to detect trouble until after service has been impaired and failure to properly isolate a faulty unit due to a shortcoming of the software and/or documentation.
- *Procedural error.* In short, these are "cockpit" or craft errors which have caused loss of service. Examples may include inputting wrong translation data or taking incorrect action during repair, growth, and update procedures.

Based on earlier studies of outage performance, an allocation of 3 minutes per year of total system downtime has been made to each of the above categories.[2]

The most important finding in the switching system outage study was that over 40 percent of outages were caused by procedural errors directly related to digital switch maintainability issues.

To reduce digital system outage, a concerted effort is required in all four categories mentioned above. However, this chapter focuses on the reduction of system outages by proper digital switch maintenance, since currently this is the highest contributing category. The next few subsections elaborate on areas that need to be studied to improve digital switching system maintainability.

7.6 Impact of Software Patches on Digital Switching System Maintainability

The frequency of generic releases for a large digital switching system is usually limited to a few times a year; however, some digital switching systems are beginning to deploy new releases more often. In between these releases all software corrections are incorporated via *patches*.[3] Patches are a "quick fix" or program modification without recompilation of the entire generic release.

In the case of real-time operational systems, it is usually difficult to install patches since the digital switching system works continuously and patches have to be applied without bringing the system down.

7.6.1 Embedded Patcher Concept

The concept of a resident patcher program for digital switches has evolved over the last 15 years or so. In first-generation digital switches, field patching was performed by hard writing encoded program instructions and data at absolute memory locations. This technique, though viable, created many problems in the operation of a digital switch. Under this hard write/read concept, mistakes were made in applying the wrong data to wrong addresses, patching incompatible generic releases, and applying patches that were out of sequence. Embedded patcher programs that operate as software maintenance programs and reside in digital switches have alleviated some of these problems.

Proper design specification of digital switching functions, coupled with exhaustive regression testing of software-hardware interfaces, could go a long way in reducing the number of patches in the field. However, the current state of digital switching software requires large numbers of patches needing excessive maintenance effort by the owners of digital switching systems.

7.7 Growth of Digital Switching System Central Offices

Most digital switching systems need to be upgraded or "grown" during their lifetimes. This process represents a major effort for maintenance organizations such as SCCs and ESACs. A digital switch may be upgraded in software or hardware, and sometimes in both. The complexity of upgrading a digital switch comes from its nonstop nature, real-time operational profile, and the complexity of software and hardware involved. The exact upgrade process for each digital switch is usually documented by the supplier, and it should be well studied to ascertain its impact on switch operation before, after, and during the upgrade process. Criteria for successful upgrades should be well documented and the results recorded after each upgrade attempt.

7.7.1 Generic Program Upgrade

The operational profile of a digital switching system requires that a minimum amount of system downtime be incurred when a new generic release is installed in an operational switch. Usually this process varies among suppliers, and a thorough analysis of this procedure is needed. The most important aspect of a generic program upgrade process is not the upgrade process itself, but how a digital switch is prepared to accept a new release. At least the following points need to be covered in the method of procedure (MOP) along with other detailed items that are specific to a digital switch and a CO:

- Time line for the entire upgrade process
- Availability of the switch during that period
- Dumping of existing data tables that need to be repackaged with the new release
- Verification of old tables with new tables to ensure that all old functionalities are supported in the new release
- The synchronization of hardware availability and software upgrade if hardware upgrade is included along with software upgrade
- Establishment of software patch levels for the upgrade process
- Supplier support before, during, and after the upgrade of the generic release

7.8 A Methodology for Reporting and Correction of Field Problems

In the digital switching environment, the internal and external (field) reporting of faults usually follows a similar scheme. A very simplified problem-reporting system is shown in Fig. 7.2. Fault reports from various sources such as testing, first office application failures, operational (CO) failures, and failures observed during the upgrade process are sent to a fault-reporting database. This database can be used to record and assign fault report numbers, fix priorities (e.g., critical, major, and minor), and track time required to fix. The formal problem report can then be captured by fault report metrics and forwarded to the module owner for correction. Depending on the type of fault, the module owner can decide to fix the problem in the current generic program with patches or to postpone it for compiled correction in the next generic program. The fault-reporting metrics can then be used to record correction history. These metrics can also be enhanced to break down the causes of failures and aid in root-cause analysis of faults.

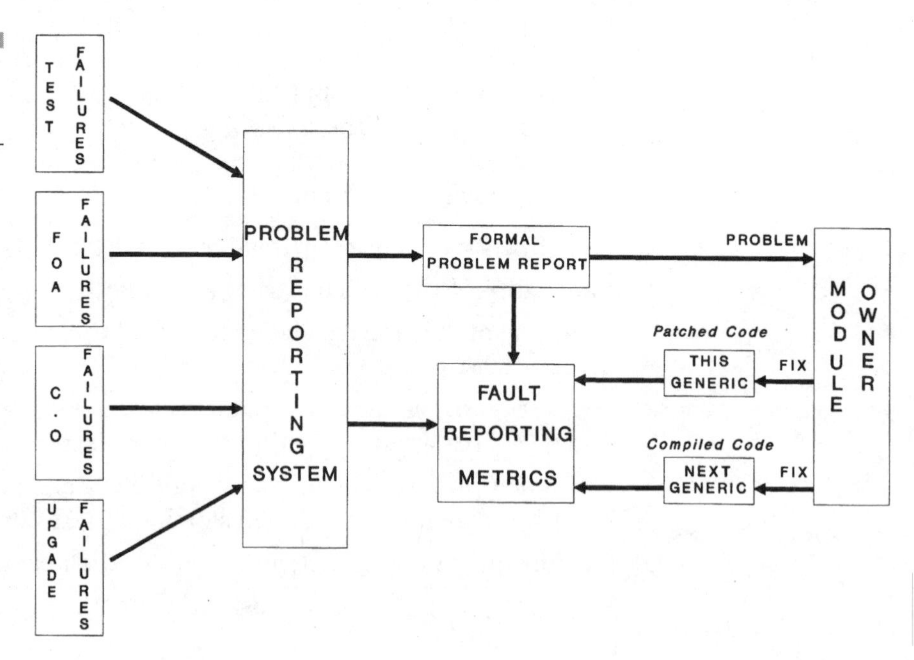

Figure 7.2 A simplified problem-reporting system

7.9 Diagnostic Capabilities for Proper Maintenance of Digital Switching Systems

Effective diagnostic programs and well-thought-out maintenance strategies play a very important role in the proper maintenance of digital switching systems with reduced maintenance cost. Most of the COs will not stock large amounts of circuit packs because of the prohibitive cost, but will use a centralized location where all types of spares are stored and maintained. Most COs are also remotely managed via SCCs and ESACs which require that the digital switching systems maintenance programs support remote diagnostics as well as provide high-accuracy diagnostic results. In some of the remote sites, it may take hours before a maintenance crew can make any circuit pack changes. In the past, switching systems employed a large number and types of circuit packs, and diagnostic capabilities were of great importance. Although modern digital switching systems are using a smaller number of circuit packs, the importance of proper diagnostics has not diminished, since a single high-density circuit pack impacts many functionalities of a digital switch. Therefore, it is imperative in the overall evaluation of a digital switch that the diagnostic capability of a switch be considered an item of high importance.

7.10 Effect of Firmware Deployment on Digital Switching Systems

The recent trend toward distributed processing in digital switching systems has resulted in increased use of firmware. The impact of firmware on digital switching system reliability and maintainability can be substantial. Most intelligent subsystems in digital switching systems require resident nonvolatile object code for the purpose of booting or bringing the system on-line after a loss of power or a system failure. These semiconductor memory types are often referred to as *firmware* devices. The term *firmware* is often used to include the program code stored in the device. For telecommunications applications, firmware can be defined as *executable code or data which are stored in semiconductor memory on a quasi-permanent basis and require physical replacement or manual intervention with external equipment for updating.*[4] With the trend toward distributed architecture in digital switching systems, the use of microprocessor controllers embedded throughout the system has increased dramatically. As a result, typical digital switching systems may have 20 to 30 percent of their program code embedded in firmware (some digital cross-connect systems and subscriber carrier systems have 100 percent of their program code embedded in firmware).

Most present-day switches incorporate many call processing functions on the line cards, and these line cards can perform many switching functions by themselves. These line cards are capable of detecting line originations, terminations, basic translation, service circuit access control, etc. Most programs which provide these functions are firmware-based. Many vendors choose this arrangement since firmware-based programs require no backup magnetic media and provide local recovery of line service with minimal manual intervention.

While the semipermanent code storage aspect of firmware provides a necessary function, it requires physical replacement or manual intervention with external equipment for updating. The updating process may involve erasing and/or programming equipment or special commands and actions from a host system for updating electrically erasable/programmable firmware devices. During the updating process, the switching system controllers may be required to operate in simplex (without redundancy). The updating process for firmware can have a significant impact on the operational reliability of a switching system, particularly if firmware changes are frequent.

7.10.1 Firmware-Software Coupling

The basic notion of "coupling" between firmware and software evolved slowly in the telecommunications industry. Telephone companies became aware of the importance of firmware in digital switches when the companies were required to change a large number of firmware packs upon the release of new software updates. The need to change significant numbers of firmware packs as part of generic updates has created a number of problems, including these:

- Increased simplex times for switches during the firmware update process
- Increased switch downtimes due to system faults while in simplex mode, required initializations for firmware changes, insertion of defective firmware circuit packs, and damaged circuit packs due to electrostatic discharge (ESD)
- Increased maintenance problems due to procedural errors
- Delays in the upgrade process because of shortages of correct versions of firmware packs
- Increased incompatibility problems between firmware and operational software

A measure of coupling between firmware and software can be established as the *ratio of firmware circuit packs,* which are changed in conjunction with a generic or major software change, *to the total number of firmware circuit packs in the system.* A low ratio indicates a "loose" coupling between firmware and software, and a high ratio indicates "tight" coupling.

An historical study of the frequency of changes and the associated ratios can be used to assess the degree of coupling between firmware and software. Industry requirements[5] seek decoupling between firmware and software as far as possible. They state, "To reduce the frequency of firmware changes in the field, firmware should be decoupled as far as possible from other software. The extent of coupling should be documented. A list of coupled firmware and the firmware's function should be provided."

7.11 Switching System Maintainability Metrics

A metric-based methodology for assessing maintainability of digital switching systems is presented next. Again, note that what is presented in this section is not meant to replace any standard way of assessing maintainability, but simply represents an initial set of metrics that need to be collected and for which appropriate scores are applied. Figure 7.3 shows a set of sample metrics with arbitrary scores.

Figure 7.3 Switching software maintenance questionnaire

SCORE → / ↓ MAINTENANCE PROCESS	0	1	2	3	4	5	SCORE
1. UPGRADE PROCESS SUCCESS RATE	40% OR LESS	50%	60%	70%	80%	90% OR MORE	
2. NUMBER OF PATCHES APPLIED PER YEAR	600 OR MORE	500	400	300	200	100 OR LESS	
3. DIAGNOSTIC RESOLUTION RATE	45% OR LESS	55%	65%	75%	85%	95% OR MORE	
4. REPORTED FAULTS CORRECTED IN DAYS (CRITICAL FAULTS)	6 OR MORE	5	4	3	2	1 OR LESS	
5. REPORTED FAULTS CORRECTED IN DAYS (MAJOR FAULTS)	55 OR MORE	50	45	40	35	30 OR LESS	
• • • • • •							

N = NUMBER OF QUESTIONS

AVERAGE SCORE = $\sum$ SCORE / N

7.11.1 Upgrade Process Success Rate

The first metric shown in Fig. 7.3 assesses the upgrade process. As an example, if the upgrade process for a digital switching system is successful only 40 percent of the time, a score of 0 is given, a score of 5 is given for success rates over 90 percent. Note that the scores shown here represent a sample guideline. A thorough understanding of the upgrade process for a particular digital switching system is necessary before one can generate scores as shown in Fig. 7.3. Important questions as to what constitutes a successful upgrade process, the impact of customer cooperation during the upgrade process, and the time required for the upgrade process need to be considered in measuring the success of an upgrade process.

7.11.2 Number of Patches Applied per Year

As mentioned earlier, a large number of patches impact digital switching system reliability and maintainability. Therefore, the number of patches applied to a system per year is a good indication of system maintainability. An arbitrary sample-scoring example is shown in Fig. 7.3. Some important questions need to be addressed. Is a patch generated for every software-correcting fault, or are a number of faults corrected with each patch? Are the CO personnel involved in screening and applying patches to their switches, or does the supplier do it automatically? These types of questions will be helpful in generating a comprehensive set of metrics for patching. Figure 7.3 shows a situation in which a single fault generates a single patch and the CO personnel are involved in patching the switch. The example shows that if the number of patches is greater than 600, then a score of 0 is entered for that particular switch; if there are 100 patches or fewer, then a score of 5 is entered; and so on. Note that these numbers are arbitrary and are not intended to replace any existing requirements.

7.11.3 Diagnostic Resolution Rate

In a modern digital switching system, it is extremely important that the diagnostic programs correctly determine the name and location of a faulty unit, down to the circuit pack level. Therefore, diagnostic programs should have good resolution rates, and this capability becomes more important when the CO is not staffed, the diagnostic is conducted remotely, and a technician is dispatched with correct circuit packs. Repair times that were used in Markov models (Chaps. 3 and 4) will depend on the accuracy of diagnostic programs. A less accurate diagnostic program will require additional diagnostic runs for identifying defective circuit packs, thus increasing repair times. An arbitrary example is shown in Fig. 7.3, which assigns a value of 0 if the diagnostic program can pinpoint defective circuit packs with an accuracy with 45 percent or less and 5 if the diagnostic accuracy is over 95 percent. Again, this is just an arbitrary example. Much detailed knowledge of the digital switch needs to be acquired before this metric can be accurately generated and used.

7.11.4 Reported Critical and Major Faults Corrected

Clearly, fault reporting and fault correction play a very important role in maintaining a digital switch. There are some strong industry guidelines in this area. For example, refer to Bellcore's *Reliability and Quality Measurements for Telecommunications Systems,*[6] which requires that all critical faults be fixed within 24 hours and all major faults in 30 days or fewer. The example shown in Fig. 7.3 establishes a score of 0 if critical faults were not corrected in 6 days or fewer and 5 if the critical faults were corrected within 1 day. Similarly, for major faults a 0 score is entered if the major faults are not corrected in 55 days or more and a score of 5 for 30 days or fewer.

7.12 A Strategy for Improving Software Quality

A strategy for improving digital switching system software quality is shown in Fig. 7.4. It is based on a process metric, defect analysis, and a continuous-improvement program. The importance of a good measurement plan cannot be overemphasized in the arena of software process improvement. A good example of software metrics is Bellcore's *In-Process Quality Metrics,*[7] and the field metric is Bellcore's *Reliability and Quality Measurements for Telecommunications Systems.*[6] These two measurement systems are used extensively in the United States by the telecommunications industry and are now being implemented in Europe. However, the methodology described here is independent of any measurement system, but depends on measurement systems that control software processes and field failures.

Let us consider this methodology in detail. Figure 7.4 shows five distinct processes. We begin at the top.

7.12.1 Program for Software Process Improvement

This represents the heart of the system. Software processes for the digital switching system are usually large, complex, and multilocational. These processes must be formalized (i.e., documented) and baselined by putting them under a configuration management system. This will allow tracking of any changes to the process and help the process administrator to better understand the impact. A process change does not always improve a process, but a continuous-improvement program (CIP) always does. The CIP strategy can vary greatly for different processes, projects, or products. The suggested strategy in this section assumes that the processes can be

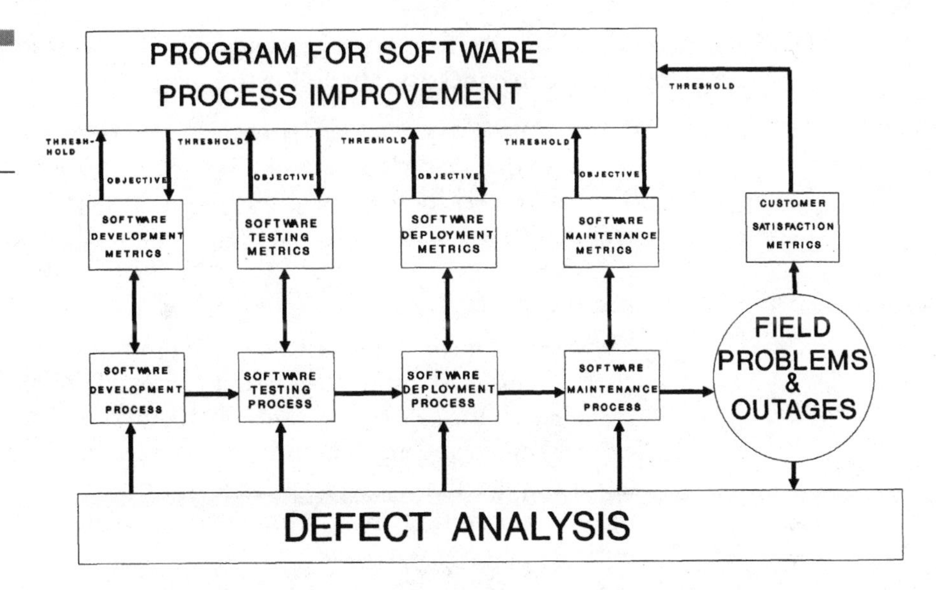

Figure 7-4 A strategy for improving software quality

instrumented. The inputs to the improvement process are the thresholds established for different metrics. These thresholds are used to observe the impact of changes on all processes. A set of new thresholds is fed to the metric system when the process is changed, enforcing tighter thresholds when required. This feedback process is implemented continuously to improve the quality of the software process.

7.12.2 Software Processes

The software processes shown in Fig. 7.4 relate to the software metrics discussed below. These include

1. Software development process
2. Software testing process
3. Software deployment process
4. Software maintenance process

7.12.3 Metrics

Figure 7.4 shows five types of metrics for this strategy. This is not to imply that a smaller or larger set of metrics will not work. Based on the needs of an organization, this strategy can be tailored to work with any set of available metrics. The following metrics are shown in the figure:

- *Software development metrics:* These metrics define measurements related to the life-cycle phases of a software development process. Typical life-cycle development phases include the software requirements process, high-level design, low-level design, and software coding. These metrics measure the effectiveness of these processes.
- *Software testing metrics:* Software testing metrics measure the effectiveness of the software testing process. Typical measurements include the number of test cases planned versus the number of cases executed, testing effectiveness, coverage, etc., applicable to all test life cycles. For digital switching systems the test life cycles can include unit testing, integration testing, feature testing, regression testing, and system testing.
- *Software deployment metrics:* These metrics are collected during the deployment of a release in the CO. The most effective metrics in this category are the application success metrics and the number of patches applied at the time of deployment. On occasion, during the application of a new release to a digital switch, the upgrade process may fail; this type of information needs to be collected to improve the upgrade process. The number of patches applied during the deployment process also must be minimized.
- *Software maintenance metrics:* These metrics are collected once the release is installed. The most important metrics are the number of software patches applied, number of defective patches found, and effectiveness of diagnostic programs.
- *Customer satisfaction metrics:* These metrics are collected from the customers of the digital switching systems. Examples are billing errors, cutoffs during conversation, slow dial tone, and other digital-switch-related problems.

7.12.4 Defect Analysis

The defect analysis is a base process for this strategy. It drives the continuous-improvement program. There are some well-defined methodologies for defect analysis,[8] and the objective here is not to define a new one. This strategy can function with any type of defect analysis methodology. After a release becomes functional in the field, it will eventually experience failures. Field failures are usually classified according to severity. Field failures that cause system outages are classified as *critical,* followed by less severe ones as *major* or *minor.* A causal analysis of all failures—especially, critical and major ones—is conducted first. After the analysis, the causes of failure are generally categorized as software, hardware, or procedural. In the next step, each failing category is expanded into subcategories. Since the strategy described here is for software processes, the hardware and procedural categories are not covered here. However, this strategy can be applied to hardware faults if the hardware development process can be mapped into life-cycle phases. Some procedural problems due to software procedures can be included in the subcategorization process.

Analysis Example Let us apply causal analysis to the hypothetical digital switching system developed for this book. Based on the software architecture of this digital switch, a software problem may have originated from

- Central processor software
- Network processor software
- Interface controller software
- Peripheral software (lines, trunks, etc.)

The next step is to identify the software subsystem that may have caused the problem:

- Operating system
- Database system
- Recovery software
- Switching software
- Application software (features, etc.)

Depending on the digital switching architecture, this subcategorization process can be long and complicated. However, once the classification of the field failures is completed and the failing software module is identified, a search is conducted to identify why this module failed and in which life-cycle phase. Usually, a patch is issued to correct the problem; however, the objective of this strategy is to fix the process so that this type of fault will not recur.

Field Trouble Report To better understand this strategy, let us analyze the following trouble report:

Name of digital switch: Digital switching system type, class 5

Location: Any Town, United States

Type of failure: Software

Duration of failure: 10 minutes

Impact: Lost all calls

Priority: Critical

Explanation: During heavy traffic period on Monday, January 1, at 9 a.m., the digital switching system lost all call processing. An automatic recovery process initialized the system. The system recovered in about 10 minutes. Yesterday night, some patches were added to correct some feature X problems. Feature X was deactivated as a precaution.

Typical Analysis This trouble report indicates a problem in feature X. Analysis of the patch and any printout during the initialization process points to the application software of the central processor. The failing module can then be identified. Further analysis of the defective module could identify the life-cycle phase by the following possibilities:

- Requirements phase (The requirement was incorrectly captured, causing the design and code to be defective.)
- Design phase (Captured requirement was correct, but the translation of requirements to design was wrong, causing defective code.)
- Code phase (Captured requirement was correct, translation of requirements to design was correct, but the written code was defective.)
- Test phase (Captured requirement was correct, translation of requirements to design was correct, written code was correct, but the testing phase did not detect the problem.)

Assume, for this particular case, that causal analysis identifies the failure as being in the test phase. Looking at the result of the causal analysis and the problem report, we see that the problem seems to be "traffic-sensitive," indicating lack of testing with high-traffic load.

A good test methodology for digital switching system software should check each feature under a realistic traffic condition before it is released. All metrics that measure the effectiveness of testing should include testing with high traffic as an input data point. The testing effectiveness threshold can now be made tighter to improve testing effectiveness. All documents related to feature testing will be changed to show enhanced traffic test requirements.

This completes the corrective feedback loop for this trouble report. Similar corrective loops need to be implemented for all trouble reports requiring process correction and improvement. This strategy enhances the software processes continuously.

7.13 Summary

This chapter covered some important areas of digital switching system maintenance. Some basic metrics were used to assess maintainability. The concept and importance of firmware coupling were introduced.

REFERENCES

1. Syed R. Ali, "Analysis of Total Outage Data for Stored Program Control Switching Systems," *IEEE Journal on Selected Areas in Communications,* vol. 4, no. 7, October 1986.
2. *Bellcore's LATA Switching System Generic Requirements* (LSSGR), TR-TSY-00064.
3. Syed R. Ali, "Software Patching in the SPC Environment and Its Impact on Switching System Reliability," *IEEE Journal on Selected Areas in Communications,* vol. 9, no. 4, May 1991.
4. Syed R. Ali, "Implementation of Firmware on SPC Switching Systems," *IEEE Journal on Selected Areas in Communications,* vol. 6, no. 8, October 1988.
5. *Bellcore's Reliability and Quality Switching Systems Generic Requirements* (RQSSGR), TR-TSY-00284, issue 2, October 1990.

6. *Bellcore's Reliability and Quality Measurements for Telecommunications Systems* (RQMS), GR-929-CORE, issue 1, December 1994.

7. *Bellcore's In-Process Quality Metrics,* GR-1315-CORE, issue 1, September 1995.

8. R. Chillarege, Inderpal S. Bhandari, Javis K. Chaar, Michael J. Halliday, Diane S. Moebus, Bonnie K. Ray, and Man-Yuen Wong, "Orthogonal Defect Classification—A Concept for In-Process Measurements," *IEEE Transactions on Software Engineering,* vol. 18, no. 11, November 1992.

CHAPTER 8

Analysis of Networked Switching Systems

8.1 Introduction

The future of the digital switching system will be highly dependent on its networking capabilities. The trend toward distributed processing will continue in conjunction with advanced intelligent networking capabilities. The next generation of digital switching systems will fully support high-speed switching modes with broadband capabilities. These concepts are explored further in this chapter.

8.2 Scope

This chapter covers the basics of networking concepts related to digital switches. Networking elements such as the service switching point (SSP), signaling transfer point (STP), and service control point (SCP) are covered. This chapter also discusses network reliability requirements, Markov models representing a hypothetical STP, and basic concepts related to the advanced intelligent network (AIN).

This chapter does not describe in intricate detail each networked element that is currently in use, but does introduce the basic function it provides. All reliability assessment techniques described so far can be tailored to analyze the networked elements as well, and this process is covered in detail. Detailed references are given for further study in this area.

8.2.1 Switching in a Networked Environment

The evolution of networking for digital switching systems has followed a remarkably similar path to that of PCs and mainframe computers. The current trend in computing is toward distributed computing connected by networks with less emphasis on central computing units. This is precisely what is happening with digital switching systems. Customers are demanding greater flexibility in the introduction of new services with short development cycles and lower costs. Typically, new releases for digital switching systems are scheduled once or twice a year. Hence, the ability to introduce new features requires the use of adjunct processors where new services could be developed and provided quickly. The concept of *intelligent network* or more recently *advanced intelligent network* has been evolving slowly since the 1970s when the concept of common-channel signaling (CCS) was introduced. This type of out-of-band signaling opened the door for external control of digital switching system functionalities. For instance, now a digital switch could send call routing information to another digital switch via a CCS link without the use of in-band signaling such as dual-tone multifrequency (DTMF). In the 1980s, digital switches were given the ability to send queries and receive responses from external databases called service control points, and this opened the door to national 800 number services. Later versions of CCS called CCS signaling 7, or CCS7, added more capabilities to linking protocols and were adapted by the International Consultative Committee for Telegraphy and Telephone (CCITT) for international use. Currently, CCS7 is the protocol for linking networked elements for digital switches.

A typical scheme for networked switching is shown in Fig. 8.1. The basic elements for such a networking scheme are as follows:

Service switching point acts as a tandem switch and can identify calls that require special handling. It formats special call requests and forwards them to a database (SCP) seeking routing information. It obtains required information by working in conjunction with STPs.

Signaling transfer point acts as a hub in a CCS7 network. It is a packet switcher for messages between CCS7 nodes and SCPs. It can also be

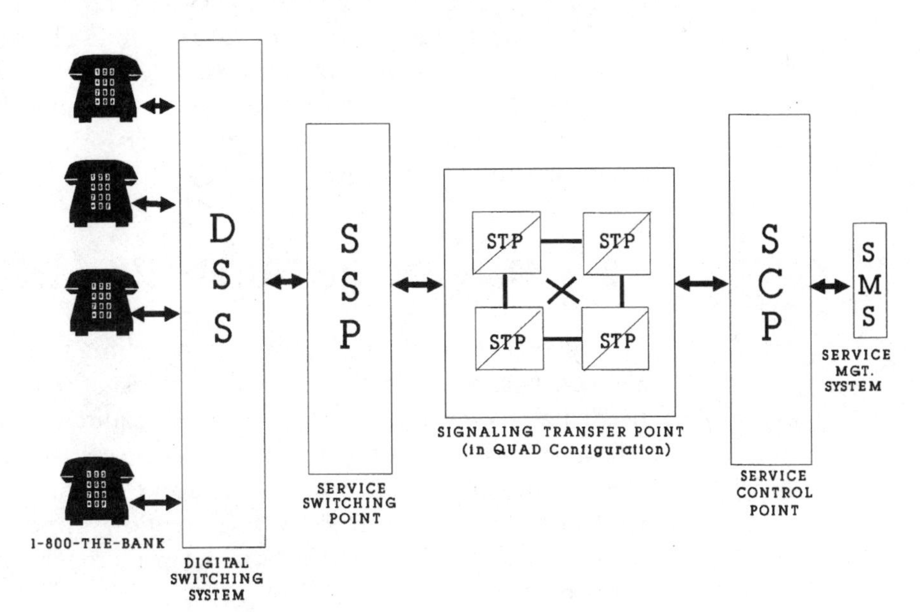

Figure 8.1 Networked switching

considered a tandem switcher for signaling messages. It has the capability of translating and routing signaling messages.

Service control point is a specialized database that can accept queries from network elements such as SSP and STP and retrieves routing information to support enhanced services, such as national 800 numbers and calling cards.

Service management system provides operational support for database (SCP) owners for managing the database and routing calls through the network that require special handling.

A basic arrangement for providing national 1-800 service using networked switching is shown in Fig. 8.1. Suppose a customer served by the local digital switching system dials a national 800 number: 1-800-THE-BANK. The digital switching system forwards this number to the SSP, and then the SSP recognizes it as a call requiring special handling and queries the SCP database through STPs that act as hubs. The response containing routing information from the SCP is passed via the STPs back to the SSP, which then completes the call through the digital switching system. In simpler terms, the data stored in the SCP associates the 800 number to a particular central office (CO) and provide the SSP with routing information along with a local directory number associated with that 800 number. This type of networked arrangement allows the 800 number to be nationally used without any danger of conflict and uses the SMS to keep it updated all the time. The advanced intelligent network uses a similar networking arrangement, but the functionalities controlled and services provided go much beyond just querying the SCP for routing information. This type of network is discussed in greater detail later in this chapter.

8.2.2 Network Reliability Requirements

The current trend toward networked switching will continue. Therefore, a much better understanding of network downtime mechanisms is needed along with associated reliability requirements for a meaningful reliability analysis. So far, downtimes associated with the digital switch itself were considered; now this concept is extended to include networked elements. Figure 8.2 shows the downtime for a typical CCS7 network with various links. This network can be divided into three functional areas:

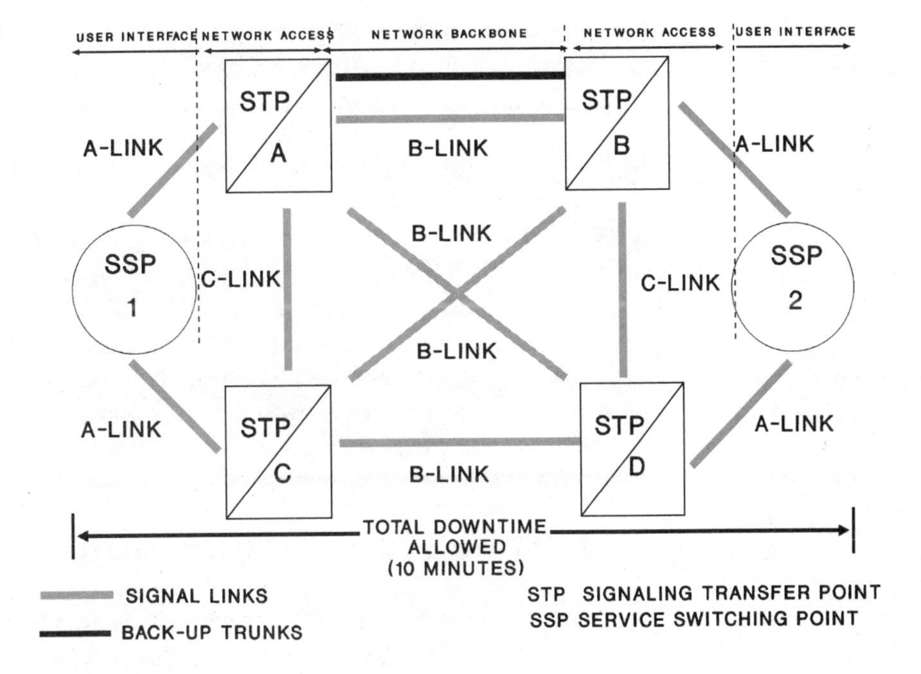

Figure 8.2 Downtimes in a typical CCS network

- *The user interface* includes SSP and a pair of A links from SSPs to STPs.
- *Network access* includes a pair of STPs connected by a pair of C links.
- *Network backbone* includes a quad arrangement of STPs connected by two pairs of B links, one of which is cross-linked. Backup trunks are also used between STPs to cover for catastrophic failures of all B links.

Network Downtimes The standard for network downtime is of extreme importance since network outages can cause regional outages of telephony service, impacting large areas of the country including international gateway connections, entire regional calling areas, airports, hospitals, and other essential services. A local outage of a digital switch is important, but the impact is usually limited to local service. Bellcore has set generic requirements for the industry in this area in TR-NWT-000246.[1] It assumes that the backbone downtime is virtually zero since it has quad redundancy, and it assigns a 3-minute downtime to the user access part and a 2-minute downtime to the network access part, which accounts for a total of 5 minutes of downtime for each end of the network. Therefore, for the arrangement shown in Fig. 8.2, the total allowed downtime is 10 minutes. Next we concentrate on analyzing a typical STP and assessing the downtime associated with it.

8.2.3 Markov Models for a Hypothetical STP

A basic reliability diagram for a hypothetical STP is shown in Fig. 8.3. It consists of a duplicated signal processor unit for processing all incoming and outgoing CCS7 messages from and to the STP. These messages are then packetized and forwarded to the duplicated packet switching unit. This model assumes that this STP unit is powered by three power supplies, of which at least two must be functional to carry the full power load; this is sometimes referred to as *2 out of 3*, or *2/3 redundancy*.

8.2.4 State Transition Diagrams

Based on the reliability diagram depicted in Fig. 8.3, three state diagrams are developed for the STP. All the methods discussed so far are now utilized in analyzing the STP model. In each case it is assumed that all the units are duplicated and that the power supply has 2/3 redundancy. Repair times are assumed to be 3 hours. For more details on the Markov chain analysis, refer to Chaps. 3 and 4.

STP Signal Processor The reliability and the state transition diagram for the STP signal processor subsystem are shown in Fig. 8.4. It assumes that the signal processor is duplicated and has five transitional states. In state 1 both processors are up and running. If one of the processors fails, then state 2, a simplex operational mode, is reached with a failure rate of λ. The failed processor can then be repaired with repair rate μ. This returns the processor subsystem to normal duplex operational mode,

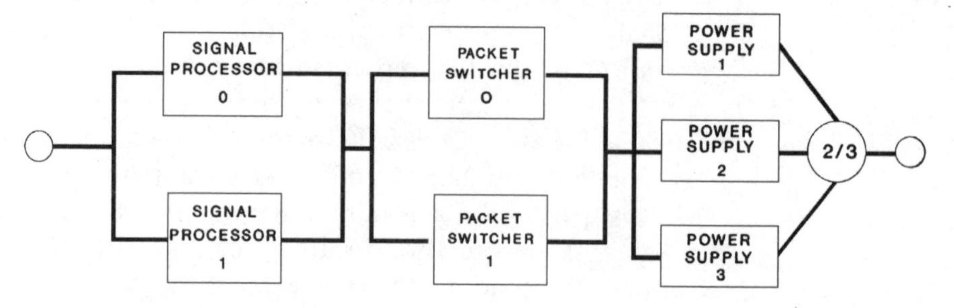

Figure 8.3 Reliability diagram for a sample STP

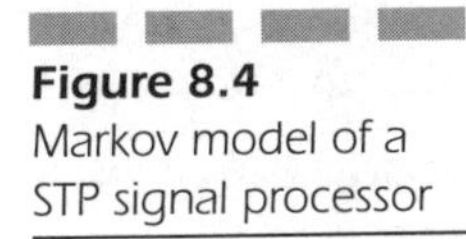

Figure 8.4
Markov model of a STP signal processor

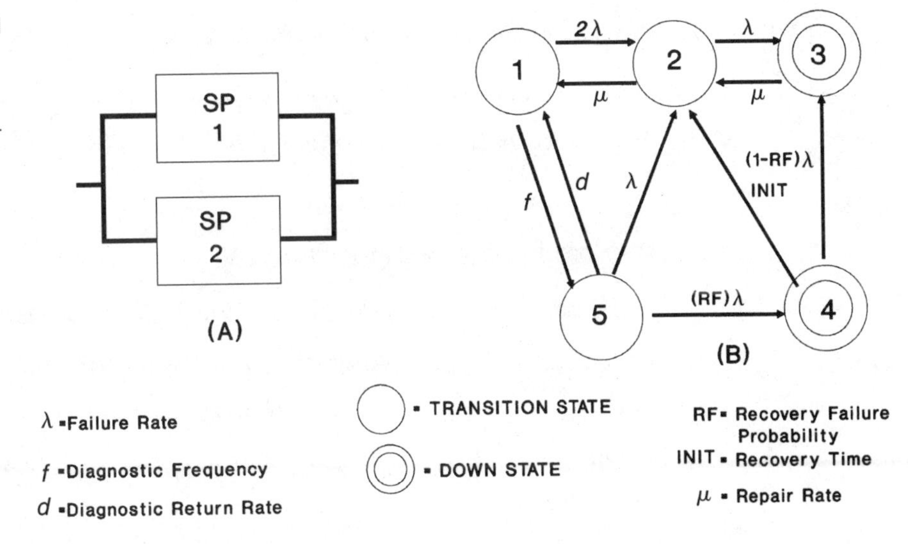

state 1. However, if the failed processor is not repaired in time and the second processor fails as well, then the processor subsystem enters duplex failure state 3. Under this condition, the processor recovery program attempts automatic recovery. State 5 is such a state; it employs automatic recovery processes and routine diagnostic programs for maintaining the processor subsystem. This attempt may or may not bring the system to a simplex operational mode, and a probability of recovery failure RF is attached to it for reaching state 4, where manual recovery procedures such as processor initialization and other techniques are used to bring the signal processor to a simplex operational state.

State Relationship Table Table 8.1 defines the parameter values and state transitions for the STP signal processor subsystem.

STP Packet Switcher The reliability and state transition diagram for the STP packet switcher are shown in Fig. 8.5. It assumes that the packet switcher is duplicated and has four transition states. State 1 depicts the normal duplex operational mode for the packet switcher. These two packet switchers are assumed to operate in parallel, one detecting any operational difficulty of the other. The probability of this occurring is assumed to be 0.50 and is termed the *detection probability* (DP). In state 4, the packet switchers are put in simplex mode, and routine diagnostic programs are executed. If the diagnostic program detects a fault in any one of the packet-switching units, it puts the packet switcher subsystem

TABLE 8.1

Modeling results for STP signal processor

Name of Model: STP Signal Processor (SP)

Total number of states = 5

Total number of failed states = 2 (failed states 3 and 4)

Parameter or Value	Meaning
$\lambda = 300{,}000$	Failure rate for all SP components in FITs (failure in 10^9 h)
$\mu = 1/3$	Repair rate, typically 3 h (1-h repair time, 2-h travel time)
$f = 1/24$	Diagnostic frequency (once in 24 h)
$d = 60/5$	Diagnostic return rate (average 5 min to complete the diagnostics)
INIT = 60/3	Automatic recovery (average 3 min)

Transitions

From State	To State	Equation	Description
1	2	2λ	Transition to simplex mode due to single SP failure (load sharing)
1	5	f	Transition to routine diagnostic state
2	1	μ	Return to duplex mode after successful repair
2	3	λ	Transition to duplex failure mode with both SP failures
3	2	μ	Return to simplex mode after successful repair during duplex failure
4	3	$(1 - \text{RF})\lambda$	Transition to state 3 after automatic recovery failure for both SPs
4	2	INIT	Transition to simplex mode after automatic recovery success
5	2	λ	Transition to simplex mode after diagnostic failure
5	1	d	Transition to duplex mode after normal diagnostics
5	4	$\text{RF}\lambda$	Transition to state 4 when diagnostic fails both SPs

Availability	0.99999
Downtime	0.854216 min/yr

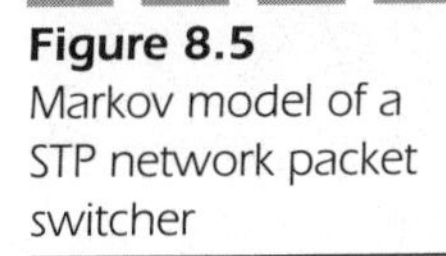

Figure 8.5
Markov model of a STP network packet switcher

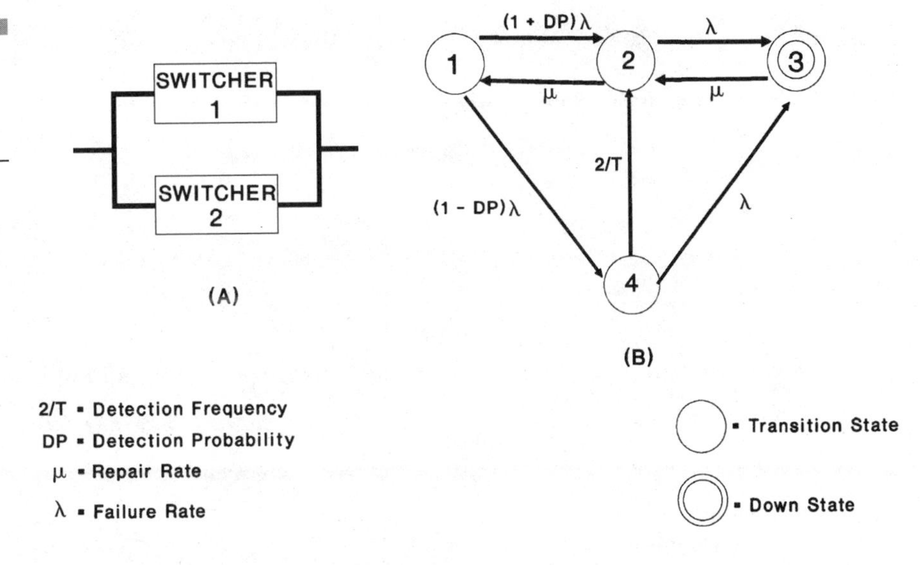

in a simplex mode, or state 2; otherwise, it returns the system to duplex mode, or state 1. State 2 represents a simplex state with one of the packet switchers marked out of service. However, when the subsystem is in simplex mode and another fault occurs, the subsystem reaches a duplex failure mode, or state 3. In all these cases it is assumed that the failure rate is represented by λ and the repair rate by μ. Once the packet switcher is repaired, it will eventually come back to its normal duplex operational mode, or state 1.

State Relationship Table Table 8.2 defines the parameter values and state transitions for the STP packet switcher subsystem.

STP Power Supply The reliability and state transition diagram for the STP power supply are shown in Fig. 8.6. It assumes that the power supply has 2/3 redundancy with three transition states. State 1 shows the normal operational mode when all three power supplies are "up" and running. State 2 is reached when one of the three power supplies fails but the system is still up and the two power supplies have enough capacity to keep the STP community up and running. The power supply subsystem reaches state 3 when any two power supplies fail. This is a down state since the failure of any two out of three power supplies will cause the STP community to go down. In all these cases, it is assumed that the failure rate is represented by λ and the repair rate by μ. It is assumed that it takes 3 h to repair a power supply. Any power supply once

TABLE 8.2

Modeling results for STP packet switcher

Name of Model: STP Packet Switcher (PSW)

Total number of states = 4

Total number of failed states = 1 (failed state 3)

Parameter or Value	Meaning
$\lambda = 200{,}000$	Failure rate for all packet switcher components in FITs (failure in 10^9 hours)
$\mu = 1/3$	Repair rate, typically 3 h (1-h repair time, 2-h travel time)
$T = 24$	Detection frequency (once in 24 hours)
DP = 0.5	Detection probability

Transitions

From State	To State	Equation	Description
1	2	$(1 + DP)\lambda$	Transition to simplex mode due to single packet switcher failure
1	4	$(1 - DP)\lambda$	Transition to duplex mode after success of both packet switchers
2	1	μ	Return to duplex mode after successful repair
2	3	λ	Transition to duplex failure mode with failure of both packet switchers
3	2	μ	Return to simplex mode after successful repair of one packet switcher
4	3	λ	Transition to state 3 after diagnostic failure of both packet switchers
4	2	$2/T$	Transition to simplex mode after diagnostic success of one packet switcher

Availability	0.99999
Downtime	0.754149 min/yr

repaired will bring the power supply subsystem back to all-up, or normal operational mode.

State Relationship Table Table 8.3 defines the parameter values and state transitions for the STP Power Supply subsystem:

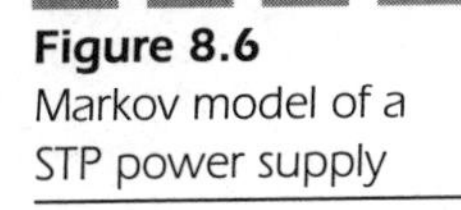
Figure 8.6
Markov model of a STP power supply

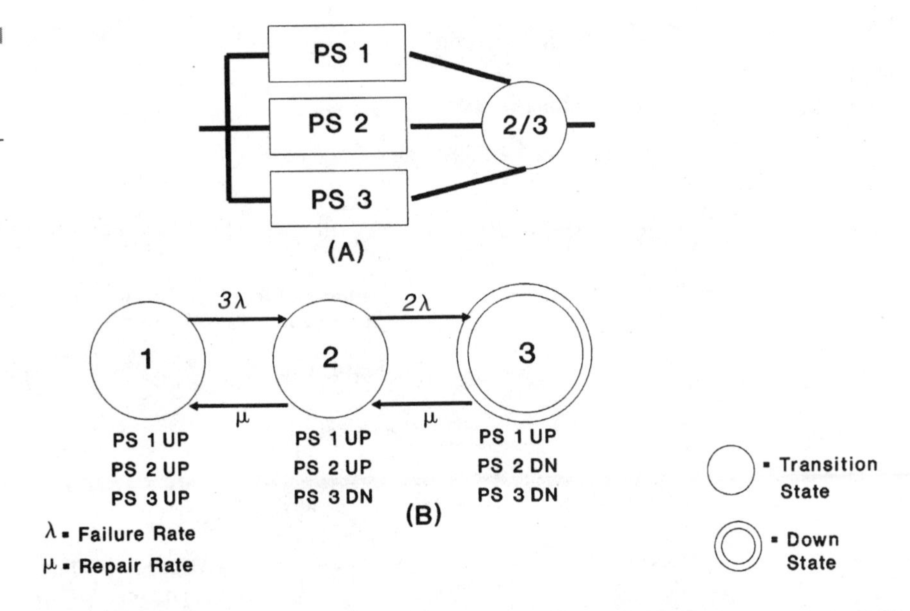

Total Downtime for the Hypothetical STP due to Hardware Failures In the last few subsections, the STP's signal processor, packet switcher, and power supply were analyzed using Markov chains. Now the total outage of STP could easily be obtained by adding the downtimes attributed to each subsection. It is shown in Table 8.4.

8.3 Dependence of New Technologies on Digital Switching Systems

In the next few sections, we introduce new telecommunications technologies that are evolving and discuss how they depend on current digital switching systems. Digital switching systems are presently the backbone of most telephony operation and will stay that way in the near future. A high-level architectural representation of these technologies will be developed so that the analysis techniques covered in this book can be easily applied to these new technologies as well. No attempt is made to fully explore each technology mentioned, since the subjects covered are vast and excellent references are available to do justice to these technologies.

TABLE 8.3

Modeling results for STP power supply

Name of Model: STP Power Supply (PS)
Total number of states = 3
Total number of failed states = 1 (failed state 3)

Parameter or Value	Meaning
$\lambda = 100{,}000$	Failure rate for all packet switcher components in FITs (failure in 10^9 hours)
$\mu = 1/3$	Repair rate, typically 3 h (1-h repair time, 2-h travel time)

Transitions

From State	To State	Equation	Description
1	2	3λ	Two-thirds normal mode, two power supplies are up and one is down
2	1	μ	Return to normal after successful repair
2	3	2λ	Transition to down state with two power supplies down and one up
3	2	μ	Return to 2/3 normal mode after successful repair of one power supply

Availability	0.99999
Downtime	0.283569 min/yr

8.3.1 Integrated Services Digital Network

A good example of new technology that is built around digital switching systems is the *integrated services digital network* (*ISDN*). This technology has evolved with digital switching systems since the early 1970s. It is based on what was then called the *integrated digital network* (*IDN*)[2] and deals with the integration of

- Switching and transmission equipment
- Voice and data communications
- Circuit switching and packet switching

As mentioned above, ISDN has evolved and is still evolving, and no attempt is made in this section to fully describe the ISDN technology.

The objective of this subsection is to introduce the reader to the importance of digital switching systems that provide circuit switching for ISDN. In very simple terms, ISDN extends the services of a telephony network to include both data and video signals along with voice. Currently, most ISDN services are provided with two types of service: basic rate and primary rate. The basic rate interface (BRI) has two B (bearer) channels and one D (delta) channel and is referred to as 2B + D. The B channels can transmit voice or data at a rate of 64,000 bits per second (b/s), or 64 kb/s, and hence the combined capacity of the two B channels is 128 kb/s. The D channel transmits signaling at 16 kb/s. A high-level conceptual diagram for ISDN components is shown in Fig. 8.7.[3] From a network analysis point of view, ISDN can be visualized as a parallel implementation of the circuit switching network, packet switching network, signaling network, and wideband switching network. All these networks are connected through the central office (CO). These services are available to users via terminal equipment (TE). Users employ the specialized TE to communicate with the CO. The transmission between the user and the CO is purely digital via the B and D channels. The CO recognizes ISDN calls and separates the B and D channels.

For a typical ISDN call, a connection request is sent from one CO to another via the D channel. As shown in Fig. 8.7, the D channel is further divided into a packet network and a signaling network. To initiate a connection, the signaling network sends a CCITT no. 7 protocol signal or a common-channel signaling (CCS) request to the other CO with particulars about a call. Once a connection is established between the two ISDN users, the B channel provides the digital speech path between users. At this stage the user is able to communicate with the other user via speech on channel B and data on channel D. With 2B + D service, users also have access to an additional B channel for data or voice.

TABLE 8.4

STP Subsystem	Downtime, min/yr
Signal processor	0.871435
Packet switcher	0.754149
Power supply	0.283569
Total	1.909153

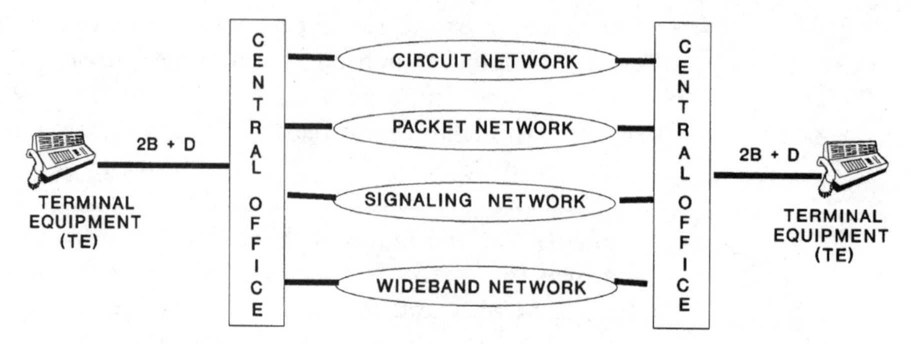

Figure 8.7
Basic ISDN concept

For analytical purposes, the wideband network could be used for wideband services such as video and Videotex in conjunction with fiber-optic connections.

All the analytical techniques developed so far can now be easily applied to ISDN, using the architecture shown in Fig. 8.7. For example, the analysis of the circuit switch is similar to that discussed in Chap. 4, the analysis of the packet network is similar to that of the packet switcher discussed in this chapter under STP analysis, and so on.

8.3.2 Current Trends in Digital Switching Systems

This subsection explores current trends in digital switching systems and presents some high-level models depicting these trends. Again, the intent here is to introduce some basic concepts that can be utilized in analyzing these platforms, and not to fully cover such developments, since they are beyond the scope of this book. However, a thorough knowledge of the evolving technologies is necessary to conduct an effective analysis of such systems.

Independent AIN Platforms High costs and long development intervals for new features in digital switching systems currently present a bottleneck in digital switching system evolution. AIN is still evolving, for the last few years, newer concepts of the advanced intelligence network are being field trailed by the telecommunications industry.[4] However, the basic concept of AIN is shown in Fig. 8.8. The hypothesized AIN switching system will be able to deploy new services to a customer quickly at lower cost and with a faster feature development cycle. The arrangement will also allow the customers to manage and upgrade existing services. The basic programming for core features will still exist in digital switching systems, but customers will have the flexibility of creating and managing their own features and services. The concepts of the intelligent peripheral (IP) and a system operations center (SOC) are also shown in Fig. 8.8. These components would be essential to support and maintain new features created for AIN. The AIN nodes discussed above could then create services without any extensive programming effort. These nodes will also allow the customers to create and maintain their own services. An SCP adjunct could also be used to provide specialized features that are not normally found in a regular SCP application. For example, it could be programmed to support a specialized database for use with cellular telephony or personal communication services (PCS).

With the advent of PCS, use of a personal line number (PLN)—each customer is assigned a personal number—will become prevalent. These numbers need to be reached regardless of the person's location or area code. This type of calling feature will require the services of an intelligent network which will have the ability to recognize such numbers and send queries to SCPs and associated cellular switching systems to locate the PLN being called.

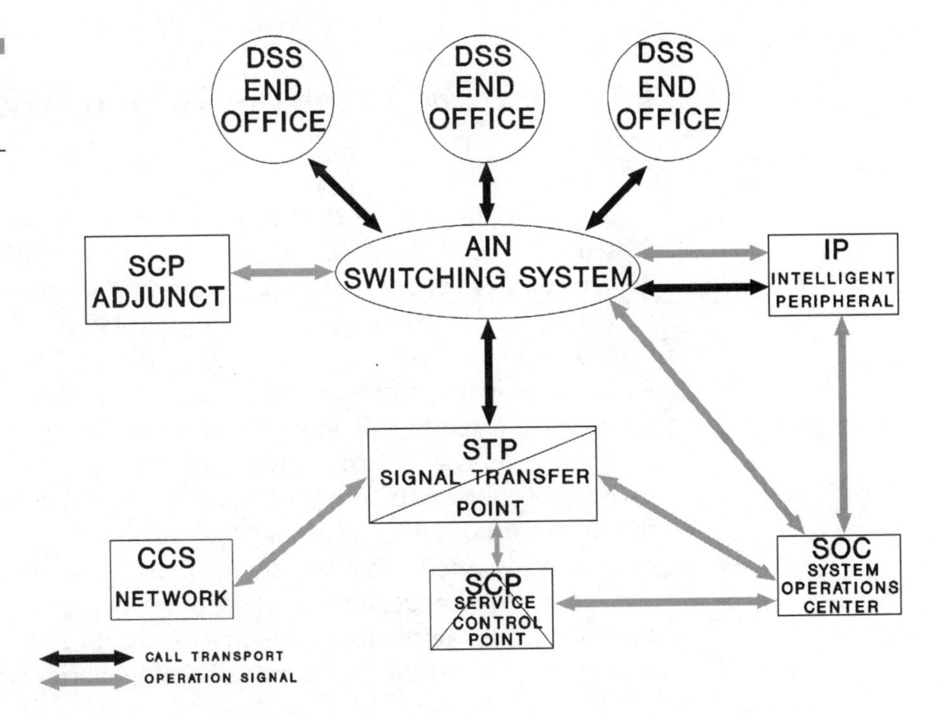

Figure 8.8 An advanced intelligent network

Example A typical call from a land line to a PCS number originates from a digital switching system end office. The digital switching system recognizes the called number as one requiring special handling. The calling information is then forwarded to the AIN switch, which will employ the SCP adjunct and will use CCS for data transfer and retrieval. Once the call is identified as a PCS call, a broadcast signal is sent to all STPs to locate the PCS customer. If the PCS customer is registered (i.e., the phone is turned on and is communicating with a cellular system, which in turn is networked to a querying STP), a temporary directory number is sent back to the AIN and consequently to the digital switching system for call connection. If the phone is turned off, a busy signal is sent to the digital switching system and no attempt is made to complete the call.

The method of analyzing such a complex system depends on the ability of the analyst to sort each segment of the network into analyzable subsystems, e.g., digital switching system, SCP, SCP adjunct, STP, IP, and so on, depending on the network itself. Each subsystem can then be analyzed separately for hardware and software reliability, as discussed in earlier chapters. These subsystems can be analyzed relative to a set of reliability requirements. Overall objectives for the entire network can also be set and tracked for compliance.

8.3.3 Future Trends in Digital Switching Systems

Digital switching system technology has evolved over the years, as discussed in Chap. 1; however, the recent trend in this evolution is more dependent on software architecture. Figure 8.9 shows such a trend. In the past, a single central processor was responsible for all digital telephony functions. The software was layered; the operating system (OS) interfaced with the switching software, which in turn controlled all application programs. Currently, this model is being replaced by a set of multiple controlling processors that control different aspects of digital switching. These processors then communicate with intelligent network elements that are autonomous and capable of making intelligent connection choices, as discussed. However, the future trend seems to lie more in the area of platform independence and the separation of basic switching functions. The switching elements in digital switching systems are bound to change from the current switching fabric technology discussed

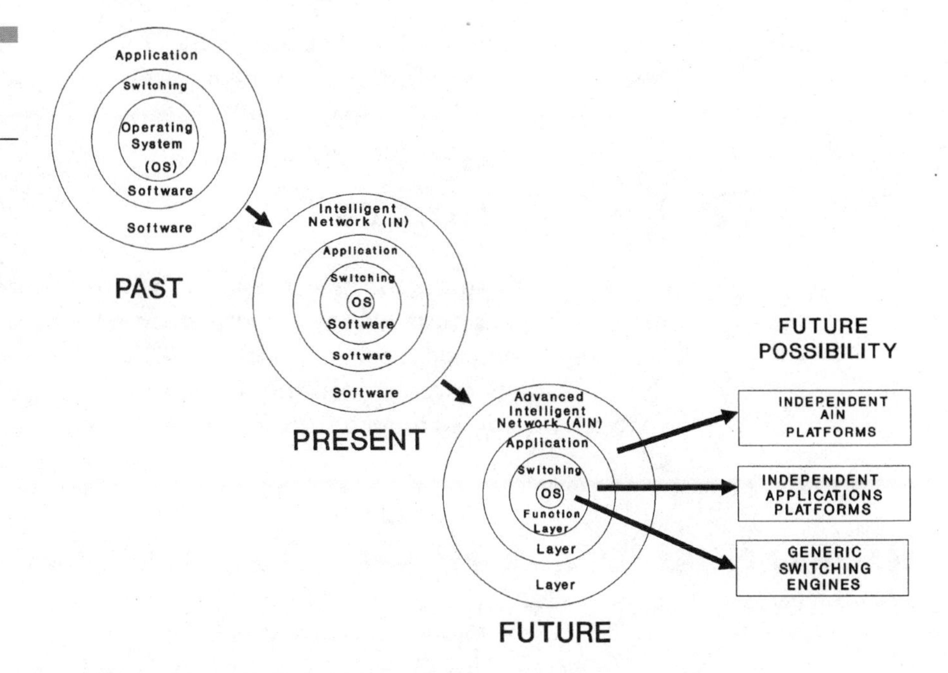

Figure 8.9 Trends in digital switching systems

in Chap. 2. New technologies that use advanced intelligent network, asynchronous transfer mode (ATM), optical elements, wireless switching, etc., will all require the separation of basic switching functionalities from application platforms. Different platforms would support different applications, and their position in the hierarchy of software implementation will depend on the application itself. For instance, implementation of AIN may need only a CCS link between AIN application and switching functions, as shown in Fig. 8.9. Other applications such as wireless switching may require trunk support at lower levels of the hierarchy.

These basic switching functions could be performed by *generic switching engines*. Most of the current digital switching systems could be reengineered to perform such functions. Indeed, international standards will be required before such generic engines can be developed or companies can develop their own internal standards to support proprietary platforms. However, this approach would avoid the possibility of a customer's choosing the generic switching engine of one manufacturer and the application platform of another.

As discussed earlier, even with the above-mentioned trends in digital switching systems, all techniques developed so far could be used once the analyst is able to understand the new architecture and group its components into analyzable subsystems. Once the subsystems are analyzed, a systematic reliability assessment of such systems can begin.

8.4 Summary

This chapter introduced the concept of a networked switching environment. A hypothetical STP was analyzed for reliability. The concepts of ISDN, AIN, and independent application platforms were discussed. It was shown that any current or future digital switching system be effectively analyzed once it is grouped into analyzable subsystems.

REFERENCES

1. *Bell Communications Research Specification of Signalling System Number 7,* TR-NWT-000246, issue 2, June 1991.
2. William Stallings, *ISDN: An Introduction,* Macmillan, New York, 1989.
3. K. T. Ma and J. Coalwell, "Integrated Circuit and Packet Switching Capabilities in the ISDN Environment," *International Switching Symposium Proceeding,* Phoenix 1987, vol. 1, IEEE Communications Society, Piscataway, NJ.
4. Donna J. Bastien, John M. Clark, and Geraldine M. Weber, "AIN: A Smarter Platform for Service," *Bellcore Exchange,* September 1993.

CHAPTER 9

A Generic Digital Switching System Model

9.1 Introduction

In the last few chapters, a reliability model for a hypothetical digital switching system was developed. Different hardware and software components of the digital switch were described, and the system was analyzed. The objective of this chapter is to go one step further—to develop a detailed model. This chapter extends the functionalities of the hypothetical digital switch to elucidate the "overall" hardware and software architectures of a typical class 5 switch. It is important for the reader to comprehend the functionalities of a digital switching system at various levels of the hardware and software architecture.

9.2 Scope

This chapter creates a generic digital switching system and its hardware and software architectures. Typical calls through the switch are traced to reveal the functionalities of an operational digital switching system. A system recovery strategy for the hypothetical digital switch and essential items in a digital switch analysis report also are covered.

9.3 Hardware Architecture

The hardware architecture of a hypothetical digital switch is shown in Fig. 9.1. This hypothetical digital switching system is based on a quasi-distributed control architecture. See Chap. 2 for details on the classification of control architectures. Note that the architecture of a working digital switching system is very complex with many subsystems. It can be covered only in a technical specification of a product, and not in a textbook of this type. The next subsections describe very high-level hardware and software architectures of a hypothetical switch. Refer to Chap. 10 for a description of some commercial digital switching systems currently deployed in the North American network.

9.3.1 Central Processor

A typical digital switching system usually employs a central processor (CP) as the primary processor, and it is always duplicated. Its function is

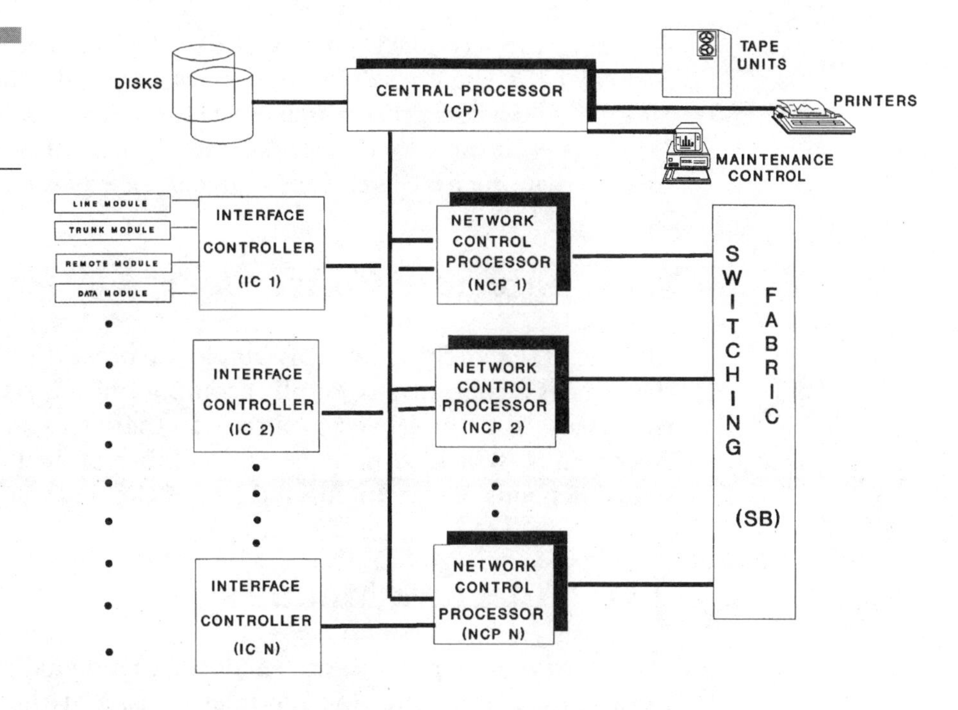

Figure 9.1 A generic switch hardware architecture

to provide systemwide control of the switching system. It usually supports secondary processors, shown in Fig. 9.1, as network control processors (NCPs). The CP usually controls high-level functions of the switch and supports operation, administration, and maintenance (OA&M) functions. When critical faults occur in the switching system, usually the CP controls the system recovery process. The CP also maintains subscriber and office data. The billing system for the switch is usually supported by the CP.

9.3.2 Network Control Processors

The network control processors are the secondary processors. Usually, their purpose is to provide call processing functions and assist in setting up a path through the switching fabric. Since most digital switching systems switch calls via a time-space-time (TST) path for call connection, this hypothetical switch is assumed to do the same. The secondary processors are usually duplicated; and depending on the desired size of the class 5 central office, a digital switch may employ a number of such processors. These processors usually interface with the interface controllers (ICs) and provide medium-level call processing support.

Generally, the secondary processors like the NCPs are associated with particular ICs. Usually a NCP keeps track of all calls that are controlled by its IC and associated paths assigned for such calls. Usually, the NCP interfaces with the CP or other NCPs to update call paths on a regular basis, so that other NCPs can get a "global" view of all calls.

9.3.3 Interface Controllers

Most digital switching systems employ a processor-based controller that acts as a concentrator of all incoming lines or trunks. These controllers use time-multiplexed output to the NCPs and provide time-switching (T switch) functions. The number of such controllers in a switch depends on the engineered size of the central office (CO).

9.3.4 Interface Modules

Different types of modules are employed in a digital switching system. Most common are the line modules (LMs) and the trunk modules (TMs). Depending on the design objectives of a digital switching system, a line module may terminate a single line or scores of lines. Most digital switching systems employ smart line cards that are processor-driven and can perform most basic call-processing functions, such as line scanning, digit collection, and call supervision. The trunk modules interface different types of trunks to the digital switching system. Most digital switching systems employ special modules to connect ISDN and other digital services to the switch. They also employ specialized module interfaces to provide enhanced services such as AIN and packet switching. The number and types of modules deployed in a digital switching system are dependent on the engineering requirements of a class 5 switch.

9.3.5 Switching Fabric

Different types of switching fabric are discussed in Chap. 2. Most digital switching systems employ at least one space or S switch. The concentrators in the ICs are usually time or T switches. The S switch is usually accessible to all NCPs. In some cases, the switching fabric is partitioned for use by different NCPs. In either case, a dynamic image of the entire network usage/idle status for the switching fabric is maintained by the CP of the digital switching system.

9.4 Software Architecture

A detailed description of digital switching system software is given in Chaps. 5 and 6. This section covers a typical organization of digital switching system software.

9.4.1 System-Level Software

Most digital switching systems employ some system-level software. Software at this level is normally a multitasking operating system (OS) and is based on a duplex mainframe computer. The function of the OS is to control each application system (AS) deployed by the digital switching system. Basic software systems for a digital switch can be classified as

- Maintenance software
- Call processing software
- Database software

9.4.2 Maintenance Software

For details on digital switching system maintenance and associated software functions, see Chap. 7.

9.4.3 Call Processing Software

Based on the architecture of the digital switching system, the call processing program can be divided into three levels:

- High level includes call processing functions that require support from a central processing unit or a central database. Examples are special feature routing, specialized billing, office data (OD), and translation data.

- Medium-level functions usually reside in the network processing units. Software supports routine call processing functions such as establishing a path through the switching fabric, verifying a subscriber, and maintaining a call map. These are referred to as *network software* in Fig. 9.2.
- Depending on the architecture of the digital switching system, many low-level functions are shared between the interface controllers and the line modules. These functions may be line scanning, digit collection, attaching service circuits, or call supervision. These are referred to as *controller* or *peripheral* software in Fig. 9.2.

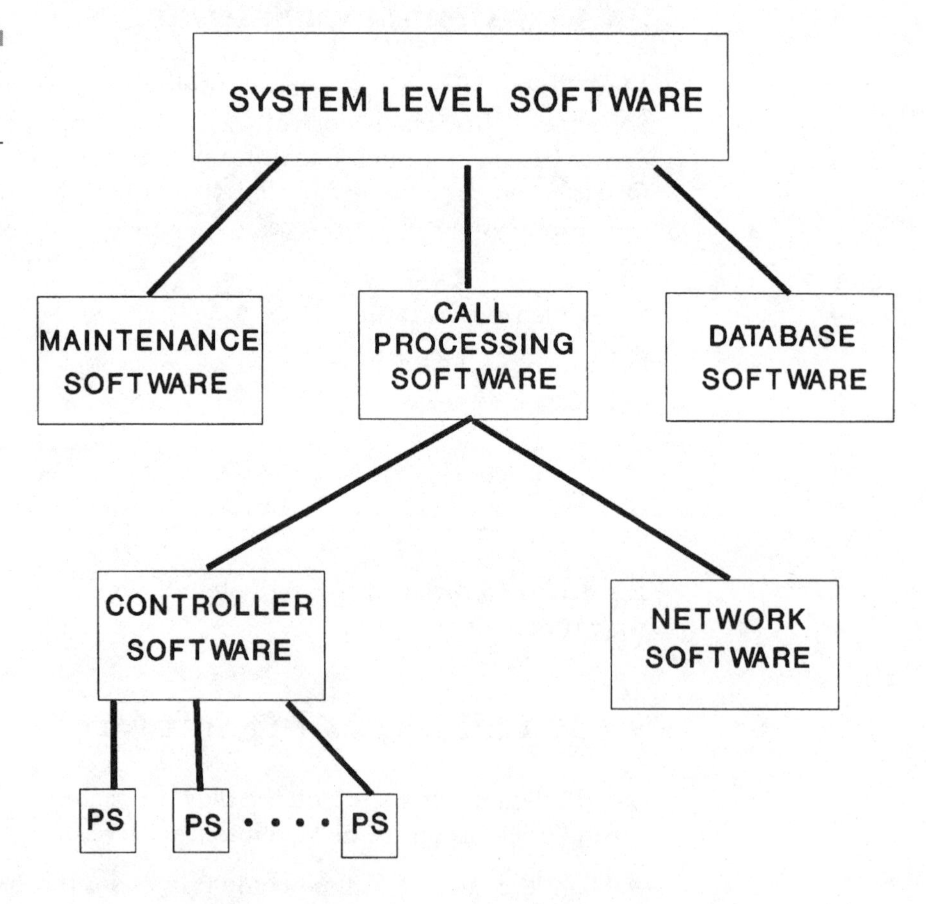

Figure 9.2 A generic switch software architecture

9.4.4 Database Software

The contents of the database software can vary greatly between digital switching systems, and within a switching system product, a switch may be engineered to provide different functions. Most digital switching systems employ a database system to record office information, system recovery parameters, system diagnostics, and billing information.

9.5 Recovery Strategy

The following is a possible recovery strategy for the hypothetical digital switching system developed for this book. An effective recovery strategy for this digital switch could be based on a three-level scheme. These schemes can be based on the three control levels developed for this digital switch:

Level 1 Initialization (INIT 1) This is considered the lowest level of initialization for the digital switch. This level of recovery initializes all components that function at level 1 control (see Chap. 2). It is controlled and directed by the interface controllers which control line modules, trunk modules, and peripheral modules (PMs).

This INIT 1 recovery could be directed specifically to initialize defined line modules, defined trunk modules, and defined peripherals. This recovery strategy selectively initializes lines, trunks, or peripherals based on the severity of the problem. This recovery can be called *local* recovery, since it can initialize peripherals locally without impacting the operation of the entire digital switching system.

EXAMPLE After a thunderstorm, a technician in a digital CO found 17 LMs, 3 TMs, and 2 PMs hung up (nonoperational). This was causing a partial outage and other operational difficulties. In this type of situation, it is appropriate for the technician to conduct a direct local recovery of these modules. Manual restoral would take too long. This type of recovery will have minimal impact on the rest of the digital switching system and will bring the digital switching system to normal operation by the low-level initialization of the digital switch.

Level 2 Initialization (INIT 2) This type of recovery can be considered as a middle-level initialization for all components that function at level 2 control (see Chap. 2). This INIT 2 recovery could be directed specifically for initializing a specified network control processor and a group of network control processors. Because of the distributed architecture of this hypothetical digital switching system, each NCP controls a number of ICs. The ICs in turn control the line, trunk, and peripheral modules. If a NCP breaks down and the backup NCP cannot switch to active mode cleanly, or if a duplex failure of a NCP pair occurs, then the operation of all ICs connected to the NCPs will be impacted. Under this condition, two types of recovery strategies need to be considered. If the problem is due to a NCP's switching from active mode to standby mode and the "switch" is not "clean," then the connected ICs may help to stabilize connections by running an INIT 1 initialization on the lines, trunks, and peripherals. If that does not help, then an INIT 2 needs to be run to initialize the NCP and associated ICs with connected LMs, TMs, and PMs. If the problem is due not to processor switching, but to a hard duplex failure in the NCP pair, then an INIT 2 needs to be run immediately. This will naturally impact all the connected ICs and associated LMs, TMs, and PMs. A multiple-NCP strategy will require initialization of a number of NCPs. Initialization of all NCPs would require a level 3 initialization, to be discussed next.

EXAMPLE

The maintenance personnel tried to switch a NCP with its redundant side after the diagnostics for the NCP failed. The NCP switch was not successful, and the digital switch lost all calls controlled by the NCP. A situation like this requires an INIT 2 initialization. This is considered a partial outage.

Level 3 Initialization (INIT 3) This is the highest level of initialization for the digital switch. This level of initialization functions at level 3 control (see Chap. 2). This INIT 3 recovery could be directed specifically for initializing the central processor (CP) and all network control processors.

This is the highest level of initialization, and it is run when the redundant CPs fail or the CP switch is not successful and the digital switching system cannot fully function with defective CPs. Under this initialization scheme, the recovery program tries to identify the problem with the last known good CP. It also seeks a "minimum" configuration for it to function. Since it cannot function fully, it will function with a reduced number of NCPs or no NCP at all, depending on the severity of the problem. Lower load or no load on the system will allow the CPs to be diagnosed effectively. Once the CP is fixed, the system will then run the INIT 2 process to synchronize all NCPs and bring them up on-line. This level of initialization will cause a total system outage.

EXAMPLE

A digital switching system starts experiencing slow dial tone, and after a time it runs an automatic INIT 2 initialization. This clears the slow-dial-tone problem, but the problem returns after a few minutes. The digital switch then starts taking repeated INIT 2's. At this stage, the technician initiates an INIT 3, which clears the problem. This type of condition usually occurs because of software corruption in the CP, and an initialization normally clears it. A thorough root-cause analysis of this type of outage needs to be conducted and the robustness of the system improved.

Manual Recovery When repeated use of INIT 3 does not recover the system, manual recovery of the digital switch becomes essential. Under manual recovery, the generic program with the last known good office data and selected subscriber data is loaded in the digital switch. Then manual diagnostics or specialized diagnostics are used to recover the digital switch. This type of manual recovery scheme is digital switch-specific, but the basic idea is as follows:

- Bring up the system with manual effort since automatic runs of INIT 1, INIT 2, and INIT 3 failed to bring the system back on-line.
- The current generic program and data may be corrupted; the system is updated with last known good generic program and data.
- Special diagnostic programs and techniques are needed to identify the problem.

The analyst should keep in mind that each digital switching system is different and may use a different strategy for system recovery. What was discussed here is just a possible strategy for this hypothetical digital switching system.

9.6 A Simple Call through a Digital Switching System

A flowchart for a typical call through a typical digital switching system is shown in Fig. 9.3. Most digital switching systems follow a similar scheme. However, note that not all digital switching systems may follow exactly the call connection sequence shown in the flowchart, but these high-level functionalities are usually covered. Details of different types of calls are given in Chap. 1, and the software linkages required during a call are covered in Chap. 5. The basic steps necessary to complete a simple call are as follows:

1. Detect off-hook condition.
2. Identify customer's line.
3. Test customer's line.
4. Provide dial tone to customer.
5. Provide digits analysis of dialed number.
6. Establish a path between the calling customer and the called customer.
7. Ring the called customer.
8. Detect answer and establish cut-through path.
9. Supervise both lines for disconnect.
10. Detect on-hook condition and disconnect.

The next few subsections describe in detail some call scenarios that typically occur in a digital switch. Indeed, many other types of call scenarios could not be covered in an effort to keep this chapter brief, and most call scenarios are switch-specific.

Line-to-Line Intra-IC Call Customer A calls customer B within the same interface controller (IC). See Fig. 9.4*a*.

When customer A goes off-hook to call customer B, the call origination request is detected by the line module. It sends a message to the interface controller which in turn sends a message to the network control processor. The NCP validates customer A's line. The interface controller attaches a digit receiver to the line and provides a dial tone to customer A. After the customer dials the first digit, the LM removes the dial tone from customer A's receiver. The dialed digits are then collected and sent to the central

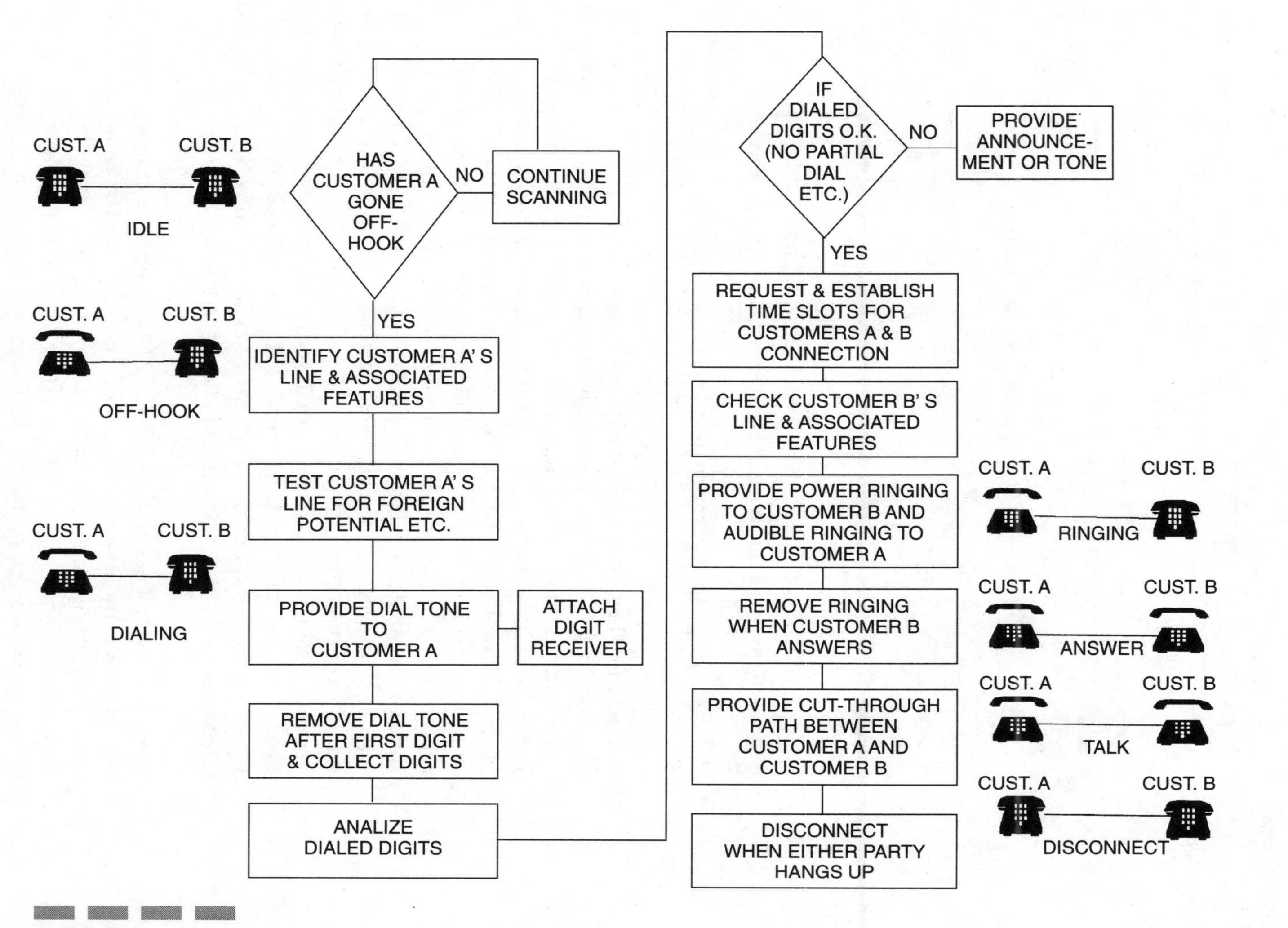

Figure 9.3
A simple call flowchart

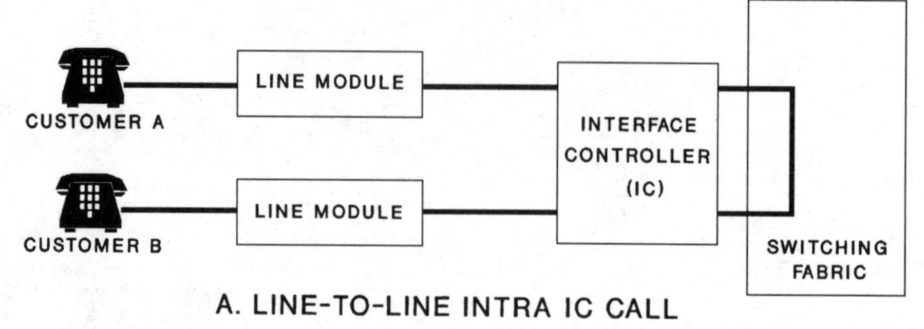

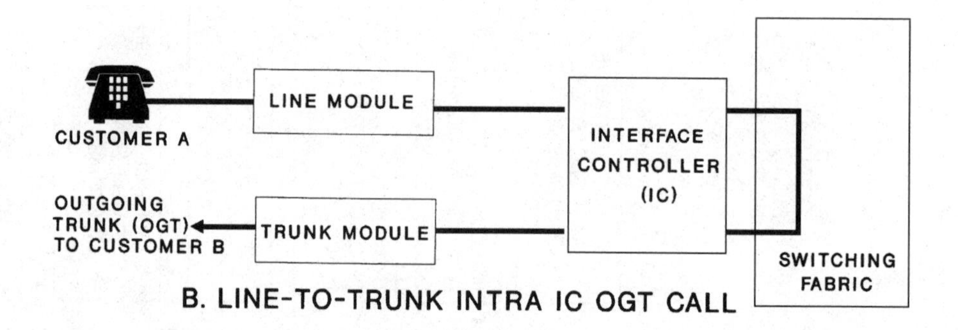

Figure 9.4
Calls within the same interface controller

processor for digit analysis. If the dialed number is valid, the NCP assigns time slots for a call connection path between customer A and customer B. If the dialed number is incorrect, for instance, has a wrong prefix, is a partial dial, etc., an announcement or a tone is given to customer A. Customer B's line is checked for busy/idle status, and a power ringing is applied to customer B's line. An audible ringing is simultaneously applied to customer A's line. When customer B answers, a cut-through path through the switching fabric is provided via previously assigned time slots. The first leg of the call from customer A uses a T switch of the interface controller, the second leg uses an S switch through the switching fabric, and the third leg to customer B uses another T switch through the interface controller. This is a typical TST connection scenario that most digital switching systems use. If either customer disconnects, the LM detects the on-hook condition and idles the connection.

Line-to-Trunk Intra-IC OGT Call Customer A calls customer B, who is served by another central office (CO), and the outgoing trunk selected lies in the same interface controller (IC). See Fig. 9.4*b*.

When customer A goes off-hook to call customer B, the call origination request is detected by the line module. It sends a message to the interface controller which in turn sends a message to the network control processor. The NCP validates customer A's line. The interface controller attaches a digit receiver to the line, and a dial tone is provided to customer A. After the first digit is dialed, the dial tone is removed from customer A's receiver. The dialed digits are then collected and sent to the central processor for digit analysis. If the dialed number is valid, the NCP assigns time slots for a path for the call between customer A and an outgoing trunk for customer B's CO or a tandem office. If the dialed number is incorrect, for instance, has a wrong prefix, an announcement or a tone is given to customer A. The terminating central office checks customer B's line for busy/idle status and applies a power ringing to customer B's line. An audible ringing is simultaneously applied to customer A's line. When customer B answers, a cut-through path through the switching fabric is provided via previously assigned time slots. As in line-to-line calls, each CO uses a TST connection. If either customer disconnects, the LM of either CO detects the on-hook condition and idles the connection. Call supervision is provided by the originating CO.

Line-to-Line Inter-IC Call Customer A calls customer B, who is located in another interface controller (IC). See Fig. 9.5*a*. This is the same as a

Figure 9.5 Calls between different interface controllers

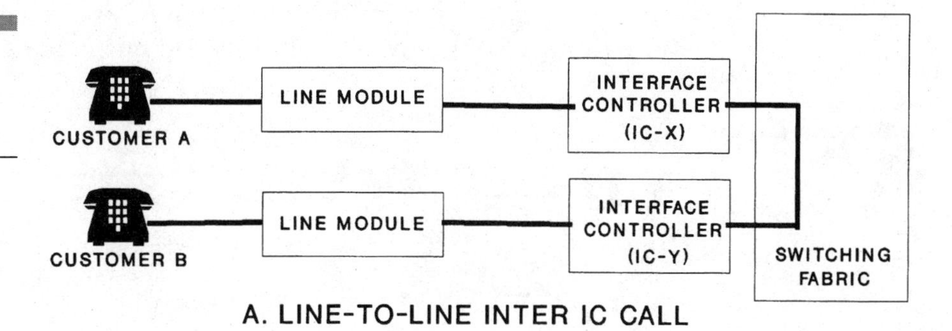

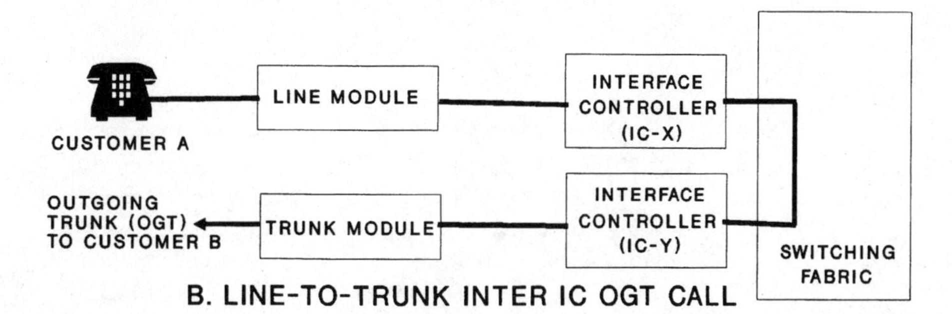

line-to-line intra-IC call, except a path through IC-X and IC-Y is established for the call. The coordination between the associated NCPs (that is, NCP-X for IC-X and NCP-Y for IC-Y) is provided by the central processor.

Line-to-Trunk Inter-IC Call Customer A calls customer B, who is located in another central office, and a different interface controller is selected. See Fig. 9.5*b*. This is the same as a line-to-trunk intra-IC OGT call, except a path through IC-X and IC-Y is established. The coordination between the associated NCPs (NCP-X for IC-X and NCP-Y for IC-Y) is provided by the central processor.

Trunk-to-Line Intra-IC IGT Call Customer A is called by customer B, who is served by another central office, and the incoming trunk selected lies in the same interface controller. See Fig. 9.6*a*.

The CO for customer B homes into customer A's CO directly or through a tandem office. It connects to customer A's CO via an incoming trunk (IGT). If the trunk and customer A's line are in the same interface controller, a path is established through the switching fabric to the line module of customer A. The associated NCP performs all time-slot assignments for the IGT and customer A's line. Line A is validated, and its idle/busy status is checked. A power ringing to customer A's line is applied by the IC, and an audible ringing is simultaneously transmitted to customer B's line via the IGT. When customer A answers, a cut-

Figure 9.6
Incoming calls to interface controllers

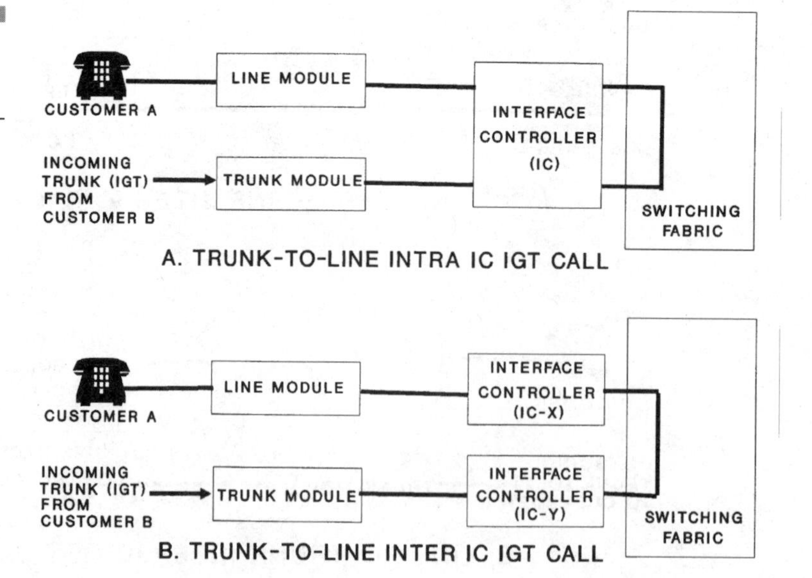

through path through the switching fabric is provided via previously assigned time slots. As in line-to-line calls, each CO uses a TST connection. If either customer disconnects, the LM of either CO detects the on-hook condition and idles the connection. Call supervision is provided by the originating CO.

Trunk-to-Line Inter-IC IGT Call Customer A is called by customer B, who is served by another central office, and the incoming trunk selected lies in a different interface controller. See Fig. 9.6*b*. This is the same as a line-to-trunk intra-IC IGT call, except a path through IC-X and IC-Y is established. The coordination between the associated NCPs (NCP-X for IC-X and NCP-Y for IC-Y) is provided by the central processor.

9.7 Some Common Characteristics of Digital Switching Systems

Most commercial digital switching systems in the North American network exhibit some common characteristics. They are described here at a high level and do not pertain to a particular switch. Chapter 10 provides some high-level details on some major digital switching systems that are currently deployed in North America.

- *Dual capability.* Most digital switching systems covered, which are primarily class 5, can also have tandem/toll or class 4 capabilities.
- *Termination capability.* Most of the large digital switching systems can terminate approximately 100,000 lines or 60,000 trunks.
- *Traffic capacity.* In a distributed environment, this depends on the digital switch configuration, and it can go as high as 2,000,000 busy-hour call attempts (BHCAs).
- *Architecture—hardware.* Most digital switching systems have a quasi-distributed hardware architecture (see Chap. 2 for definitions), since they all maintain control of the switching functions through an intermediate processor. All digital switching systems employ multiple-processor subsystems.
- *Architecture—software.* Most digital switching systems maintain a modular software design, sometimes through layering or through functionalities. They have operating systems under which application systems function. They all support database systems for office records, subscriber records, administration records, etc. They all have maintenance subsystems that support diagnostic and

switch maintenance processes. They also support billing systems for subscribers such as the automatic messaging system.

- *Switching fabric.* Most digital switching systems utilize time-space-time (TST) mode for switching calls.
- *Remote operation.* Most digital switching systems have remote switching modules (RSMs) to support switching functions in a remote location. And most remote switching systems have stand-alone capabilities, so if the main switching system (host) goes down, the remote units can still switch local calls.
- *Advanced feature support.* Most digital switching systems can support advanced features such as ISDN, STP, SCP, and AIN.

The telecommunications market is now demanding a marriage between telephony and cable television applications. This would change the nature of class 5 COs and would require broadband switching. The use of Internet around the world is placing very high demand on class 5 CO provisioning requirements. Many Internet users now connect to their Internet providers through class 5 COs and keep the connection up for long periods. Most COs were not designed for such use. New types of interfaces may be required to identify such calls and route them through designated COs equipped to handle high holding times. The cost of provisioning COs will rise since most of the cost associated with equipping a class 5 digital switching system comes from customer interfaces such as line modules, trunk modules, and service circuits. The integration of voice, data, and full-motion video as required by the Internet and other services will need to be switched through a class 5 digital switching system. The use of ATM and optical links using SONET will dominate the switching markets of the future.

9.8 Analysis Report

The analysis report of a digital switching should at least contain the following sections.

9.8.1 System Description

This section gives a high-level description of the digital switch being analyzed, with emphasis on:

- *System overview:* Describe system-level functional blocks of the digital switch.
- *Capacity:* Cover busy-hour call attempts of the digital switch for desired configurations.
- *Hardware description:* Give a detailed description of all important hardware components of the digital switch required for desired configuration of equipment.
- *Software description:* Describe the main software architecture of the digital switch with all major software components identified.
- *Call processing:* Describe the flow of different types of calls through the digital switch.
- *Features list:* Describe all base features (included with generic) and optional features that need to be obtained separately.
- *System recovery strategy:* Describe different levels of system initialization and typical times for system recovery for each level of initialization.

9.8.2 Operation, Administration, and Maintenance

This section gives a brief description of all maintenance features of a digital switch, with emphasis on:

- *Database management:* Describe all databases that need to be managed, e.g., office database, translation database, and billing database.
- *OSS interfaces:* Describe all types of operational support system interfaces.

9.8.3 Reliability Analysis

This section is the most important, and it gives a brief description of the reliability models of the digital switch and includes overall reliability findings covering

- Component failure rates: Describe the component failure rates for different circuit packs used in the digital switch.
- System reliability: Describe the results of hardware modeling of various subsystems of the digital switch.

- Software reliability analysis: Describe the results of the software analysis of the digital switching system software.

9.8.4 Product Support

This section describes the organizational structure and commitment of the organization to support the digital switching system after it is sold.

- *Technical assistance:* Describe different levels of technical support that the digital switching supplier provides and the escalation process and time limits within which the supplier will correct the fault.
- *Documentation:* List all documents that will be supplied to maintain the digital switching system and how often it will be updated.
- *Fault reporting system:* Describe a fault-reporting system that tracks all faults discovered by the operator of the digital switching system.
- *Training:* List all training courses available for telephone company personnel who will use and maintain the digital switching system.

9.9 Summary

This chapter developed and enhanced a hypothetical digital switching system to clarify hardware and software architectures of a typical CO switch. Simple line-to-line calls along with other types of calls were traced through a hypothetical CO. This chapter also described a recovery strategy for a hypothetical digital switching system and listed the essential items in a digital switching analysis report. The last section of this book discusses some common characteristics and future trends in digital switching systems.

CHAPTER 10

Major Digital Switching Systems in the North American Network

10.1 Purpose

This chapter covers most of the major digital switching systems currently deployed in the North American network. The objective of this chapter is to introduce the high-level design of different digital switching systems, not to render an analysis. The information presented here has been gleaned from publicly available publications, and consequently it may not be current. All digital switching systems constantly undergo architectural changes to keep up with new demands of the telecommunications industry. Information presented in this chapter is intended to provide a high-level view of hardware and software architectures of digital switching systems and should not be viewed as "specification sheets" for these products. To further emphasize this point, a specific number of modules, line concentration ratios, processor types, etc., are not mentioned.

It should be noted that this chapter covers most of the digital switching systems that are currently deployed in the North American network. To keep this appendix brief some digital switching systems could not be included.

10.1.1 Lucent Technologies' 5ESS Switching System*

System Overview The 5ESS switching system can be configured as a medium-size or a large class 5 digital switch, and it provides local, toll, and operator services. It was first put in service in 1982. It supports POT and Centrex services as well as advanced services such as ISDN, STP, SCP, and AIN.

Its architecture can be classified as quasi-distributed since it maintains control of the system via an administrative processor. Its switching fabric is time-multiplexed and is classified as a time-space-time (TST) network.

Hardware Architecture[1] A very high-level description of the 5ESS switching system is shown in Fig. 10.1. Three main components of the 5ESS switch are the administrative module (AM), communications module (CM), and switching module (SM). The next few sections describe their basic functions and the functions of some subcomponents.

* ESS is a registered trademark of Lucent Technologies.

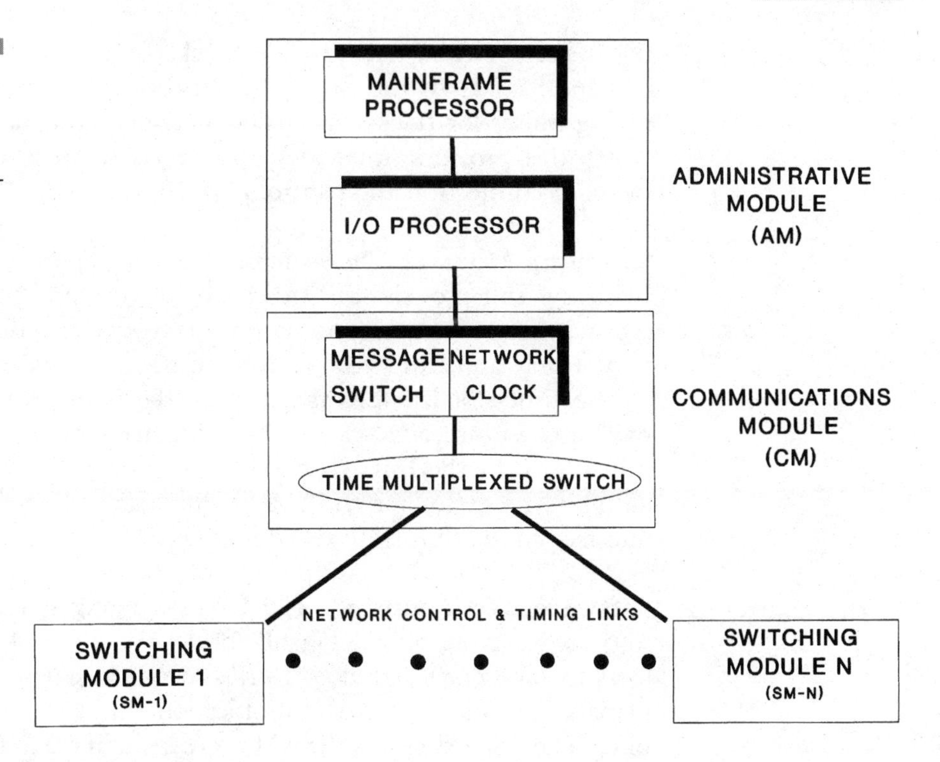

Figure 10.1 Lucent Technologies' 5ESS switch hardware architecture

Administrative Module The administrative module is based on a duplexed AT&T mainframe computer. The two processors of the AM operate in an active/standby mode; i.e., one processor is always active, and the standby processor can take over the operation of the active processor if a fault occurs in the active processor. This arrangement provides higher reliability owing to redundancy. The function of the AM is to assist in call processing functions, system maintenance, software recovery, error detection, and system initialization. It supports hard disk access and maintains system software, office recovery data, and billing data. It also interfaces with I/O processors that control video displays, printer's tape units, and the master control center (MCC). The MCC provides system status and manual control of the 5ESS switch. In the area of call processing, AM supports routing of calls to a particular switching module and tracks their availability. It also controls and allocates time slots for the time-multiplexed switch (TMS).

Communications Module The communications module (CM) provides communication between the AM and the SMs. It also provides communications between SMs via the network control and timing (NCT) links that use fiber-optic cables. The basic components of CM are the message switch that provides packet-switching functions between CM and its SMs via the NCTs; the network clock that provides synchronizing pulses for the time-division network; and the time-multiplexed switch that provides timeshared space-division switching in the CM for switched connection and control between the SMs.

Switching Module The switching module represents the first stage of switching and provides call processing. It connects lines and trunks to the switching system via interface units (IUs). Depending on the application of lines and trunks, SMs can be configured differently. The function of the time-slot interchange unit (TSIU) is to provide time-division switching within each SM. The NCT linkage between the SM and CM is time-slotted by the TSIU via the TMS. The module control unit that contains the processor for the SMs is duplicated. The TSIU and the link interface circuits (LICs) are also duplicated.

Software Architecture[2] The 5ESS switching system uses a modular software structure; refer to Fig. 10.2 for details. Its operating system, referred to as the *operational system for distributed switching* (*OSDS*), provides process management, interprocess communication, timing services, and task scheduling. The OSDS supports the AM processor and the SM processors.

Most of the software that performs administrative functions at system level resides in the AM. Some features or options that are not frequently required may also reside in the AM. The TMS software also resides in the AM.

The SM contains all programs necessary for the control of switching periphery. Actions like port control for lines and trunks, allocation of switching network paths and service units, setting and releasing the call path, scanning of lines and trunks, and outpulsing are all performed by the SM.

The function of the routing and terminal module is to support routing and screening functions for other modules, e.g., feature control that may require additional information such as call destination based on dialed digits, routing path, etc. This module also keeps status information of lines, trunks, and terminals associated with other subsystems.

The feature control module receives dialed digits and call supervision information from switching peripheral control and interprets them; and if needed, it interfaces with the database management system for additional information. It then sends messages to the peripheral control to complete any required action.

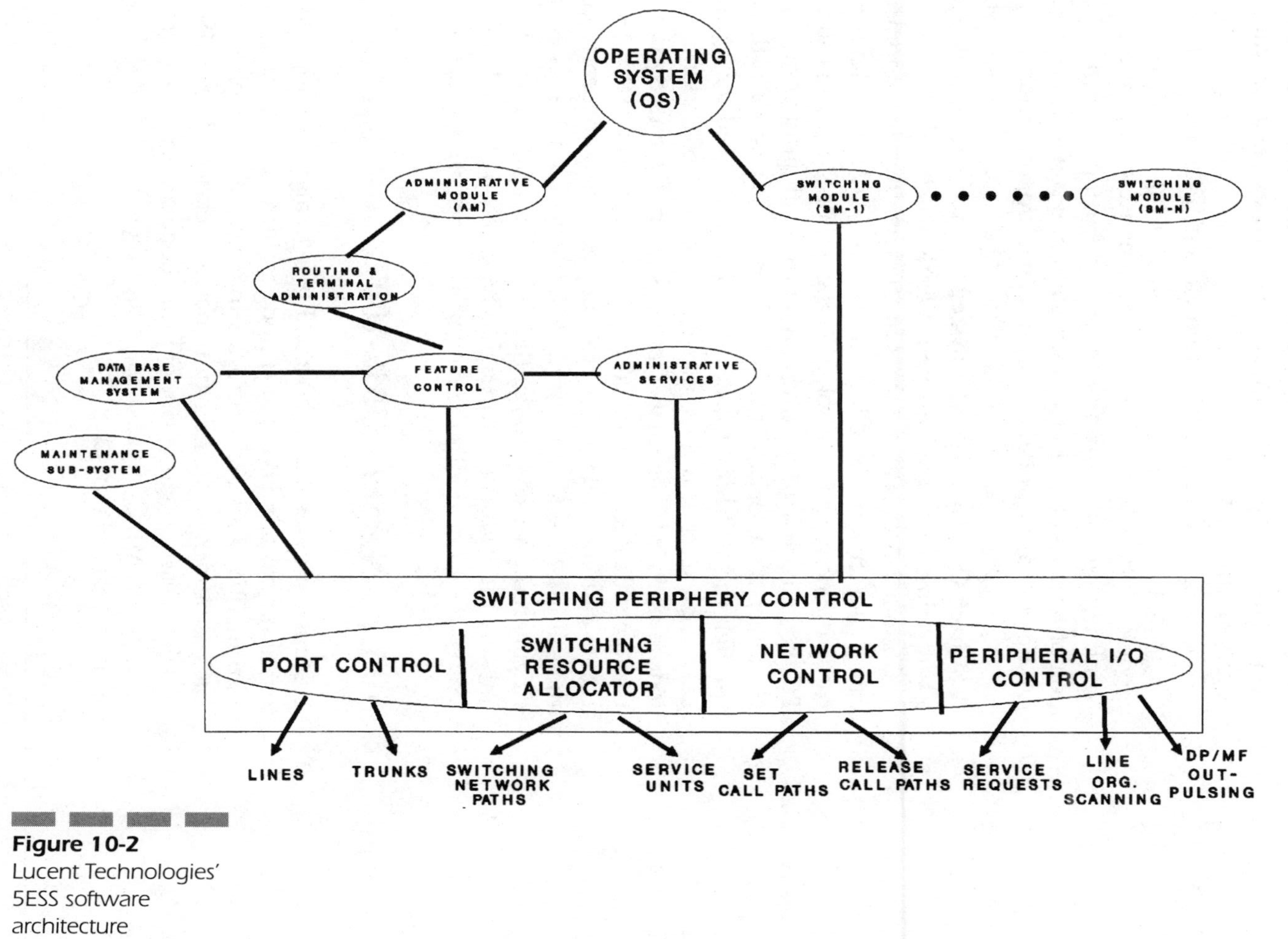

Figure 10-2 Lucent Technologies' 5ESS software architecture

The administrative services module provides billing, traffic measurements, plant measurements, and outputting network management functions. This module also interfaces with external memory and data administrative services.

The maintenance subsystem provides maintenance functions through the master control center and trunk and line workstation (TLWS). This module support line and trunk maintenance such as subscriber line and trunk testing and transmission testing. It supports removal and restoration of lines and trunks from service. It also supports other activities that can be provided through the operational support systems.

A Simple Call through the System A very simplified explanation of a simple line-to-line call through a 5ESS switch follows:

Customer A calls customer B in the same switch. Customer A in SM 1 calls customer B in SM 3. The off-hook condition of customer A is detected by the peripheral control (PC) program of SM 1. To establish a call, the PC program calls the routing and terminal administration (RTA) program. Customer A's associated features are identified, and a path is established through the switching fabric after the line is tested for foreign potential, etc. A digit receiver is then attached to the line of customer A, and a dial tone is provided. The digits from customer A are received and analyzed. Customer B's line is also tested, and a request is sent to the AM for connection to customer B. The RTA program again helps to identify and establish the call to SM 3. A path through the switching fabric is established, and ringing is connected through the TSI of both SMs. When the called party answers, the path is cut through. The called party's SM controls supervision of the call and idles the path when either party hangs up.

10.1.2 Nortel's DMS-100*

System Overview The DMS-100 switch (digital multiplexing system) is based on Nortel's DMS SuperNode architecture. It was first put in service in 1979. This architecture supports many telephony applications such as tandem switching and operator services. The DMS-100 is a class 5 switch that supports plain old telephone service (POTS), meridian digital centrex (MDC), integrated service digital network (ISDN), and automatic call distributor (ACD).

* DMS is a registered trademark of Nortel.

Its architecture can be classified as quasi-distributed since it maintains control of the system via a central processor. Its switching fabric is time-multiplexed and is classified as a time-space-time (TST) network.

Hardware Architecture[3] A high level of the hardware architecture of the DMS-100 switch is shown in Fig. 10.3. The main components of this switching system are

- DMS core processor
- File processor
- Application processor
- Link peripheral processor
- Network switching (switching fabric)
- Peripheral interfaces

DMS Core Processor. The DMS core processor consists of a duplicated set of central processing units. It functions as a high-level call handler for

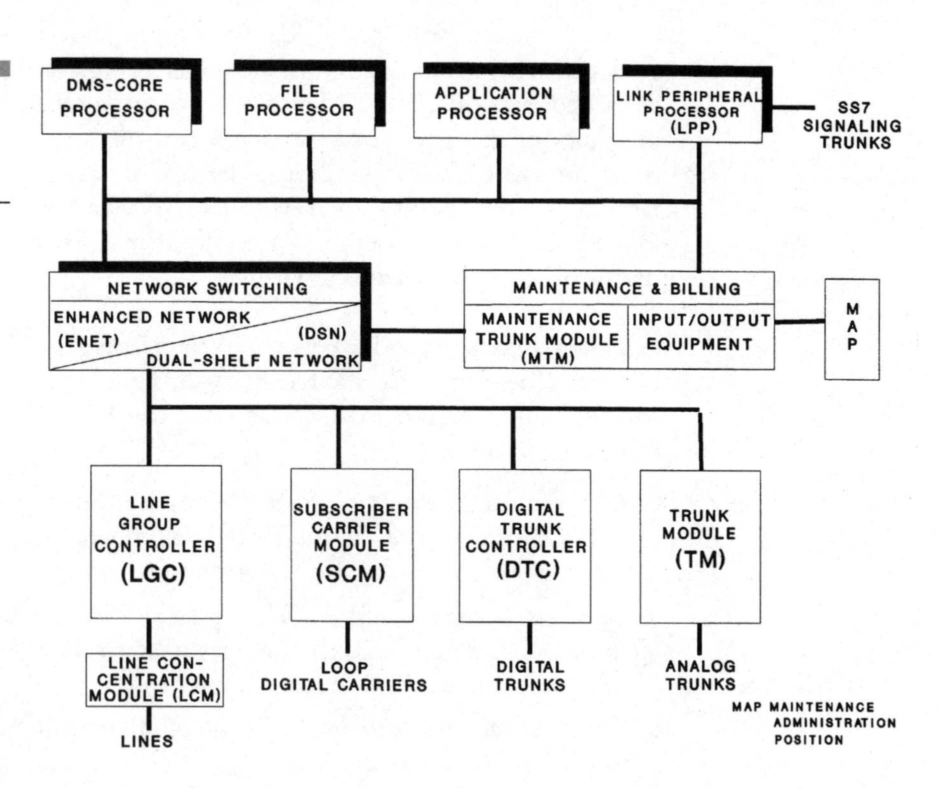

Figure 10.3 Nortel DMS-100 switch hardware architecture

the DMS-100 digital switching system and supports other peripheral controllers during routine call processing. The DMS core units also perform system control functions, system maintenance, and downloading of DMS SuperNode software.

File Processors and Applications Processors. The file processors are duplicated processing units that provide access to large data files for applications such as SCP and AMA. The applications processors are intended for specialized applications. These processors are also duplicated and allow telephone operating companies to add their own programming for an application. For instance, they can be used to provide any special operations, administration, and maintenance (OA&M) functions.

Link Peripheral Processor. The link peripheral processor (LPP) supports CCS7 and advanced data applications. It can be configured to provide different services by employing different interface units. It supports SSP, STP, SCP, ISDN, and other services. The message switch unit of the LPP is duplicated to increase reliability.

Network Switching. The DMS-100 digital switching system can be equipped with two types of switching matrix: dual-shelf network (DSN) or enhanced network (ENET). The DSN switching fabric uses time-division multiplexing to connect to the peripheral modules. It supports narrowband voice and data applications. The ENET is Nortel's newer switching fabric and is designed for the DMS SuperNode architecture. It is duplicated for reliability and is based on single-stage time switching. It can support narrowband and wideband services.

Peripheral Interfaces. The peripheral interfaces provide interconnection between lines and trunks to the network switching elements and processors. Principal peripheral interfaces for the DMS-100 digital switching system are as follows:

- *Line group controller (LGC)*: It performs medium-level processing tasks for subscriber line interfaces. It interfaces line modules with the switching fabric. It also serves remote line modules.
- *Line concentrating module (LCM)*: It provides an interface between the subscriber lines and the line group controller. Both of these interfaces provide line concentration.
- *Subscriber carrier module (SCM)*: It interconnects digital loop carriers to the switching fabric.

- *Digital trunk controller* (*DTC*): It provides digital trunk interconnection between DMS-100 digital switching system and other central offices.
- *Trunk module* (*TM*): It provides an interface between external analog trunk facilities and the DMS-100 digital switching system.
- *Line trunk controller* (*LTC*): It combines the functions of the LGC and DTC (not shown).

Maintenance and Billing. The operation, administration, and maintenance for the DMS-100 digital switching system is provided through the maintenance and administrative position (MAP). It supports activities such as system access, service orders, data recording, line and trunk testing, and alarms. The billing subsystem supports automatic message accounting (AMA). It records AMA data on tape and disk. The AMA data can also be transferred automatically to the accounting offices.

Software Architecture[4] The DMS SuperNode software architecture is shown in Fig. 10.4. The software is arranged in four layers:

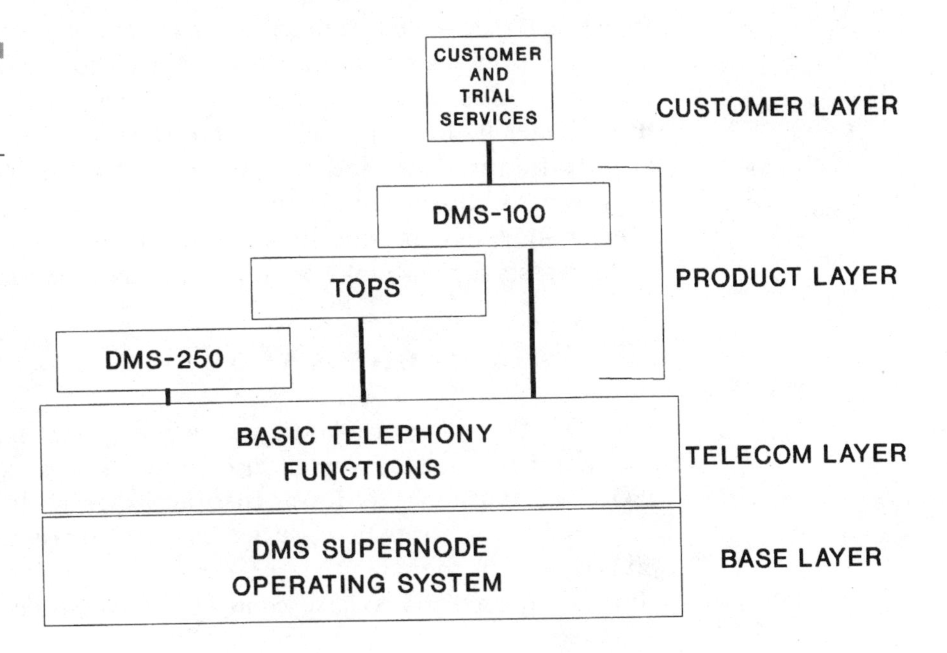

Figure 10.4 Nortel DMS-100 software architecture

- *Base layer:* It contains DMS SuperNode operating system.
- *Telecom layer:* It supports all basic telecommunications functions.
- *Product layer:* It supports different products that work under the DMS SuperNode architecture. Some of the other products are shown in Fig. 10.2; it includes the DMS-100 digital switching system as one of the products.
- *Customer layer:* This layer supports custom software and trial services specific to a customer.

A Simple Call through the System A very simplified explanation of a simple line-to-line call through a DMS-100 switch follows:

Customer A calls customer B in the same switch. When customer A goes off-hook to call customer B, the call origination request is detected by the line concentrating module. It sends a message to DMS core which validates customer A's line. A digit receiver is attached to the line, and a dial tone is provided to customer A. The dialed number is received by the LCM, and the dial tone is removed from customer A's receiver. The dialed digits are passed to DMS core for analysis. If the dialed number is valid, time slots are assigned for customer A and customer B. If the dialed number is incorrect, for instance, has a wrong prefix, an announcement or a tone is given to customer A. Customer B's line is checked for busy/idle status, and a power ringing is applied to customer B's line when it is found to be idle. An audible ringing is simultaneously applied to customer A's line. When customer B answers, a cut-through path is provided through the network via previously assigned time slots. If either customer disconnects, the LCM detects the on-hook condition and idles the connection.

10.1.3 Ericsson's AXE 10*

The AXE 10 is a large class 5 digital switching system. It provides basic POTS and business services. It also supports advance services such as ISDN, STP, SCP, and AIN. It was first put in service in 1978.

Its architecture can be classified as quasi-distributed since it maintains control of the system via a central processor. Its switching fabric is time-multiplexed and is classified as a time-space-time (TST) network.

---

* AXE is a registered trademark of Ericsson.

Hardware Architecture[5] The hardware architecture of the AXE 10 switch is shown in Fig. 10.5. It consists of the following subsystems:

- Subscriber switching
- Group switching
- Trunk and signaling
- Regional processors
- Central processor
- Maintenance
- Input/output

Subscriber Switching Subsystem (SSS). It is a subscriber line switching subsystem. Subscriber lines connect to it through line switch modules. It also acts as a line concentrator.

Group Switching Subsystem (GSS). The group switching subsystem provides time-space-time switching for incoming lines from SSS and trunks from TSS (see below).

Trunk and Signaling Subsystem (TSS). The trunk and signaling subsystem connects analog and digital trunks to the GSS.

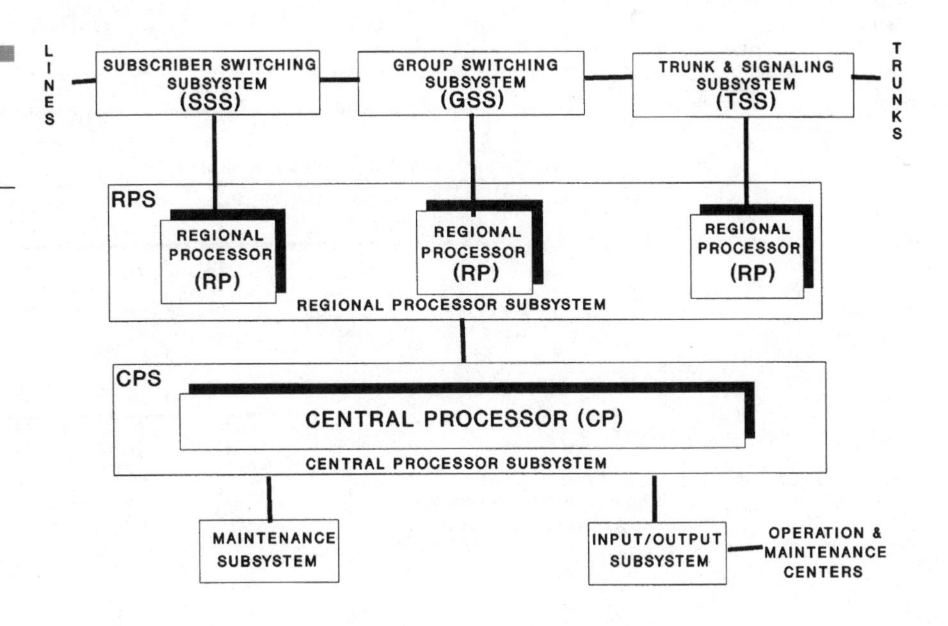

Figure 10.5 Ericsson AXE 10 switch hardware architecture

Regional Processor Subsystem (RPS). The regional processors perform routine tasks such as scanning line circuits and connect to the central processor and the switching fabric. The regional processors are duplicated for reliability.

Central Processor Subsystem (CPS). The central processor subsystem represents the main processing unit of the AXE 10 switch. It performs all complex and infrequent tasks. It also controls the maintenance subsystem and the input/output subsystems. It is duplicated for reliability.

Maintenance Subsystem. The maintenance subsystem provides all maintenance functions for the AXE 10 switch such as fault isolation.

Input/Output Subsystem (IOS). The input/output subsystem provides interconnection to operation and maintenance facilities.

Software Architecture[6] The software architecture of the AXE 10 switch is shown in Fig. 10.6. The AXE 10 software is subdivided by hierarchy:

- AXE system software: software that is common to the entire system
- AXE subsystem software: software that is common to subsystems

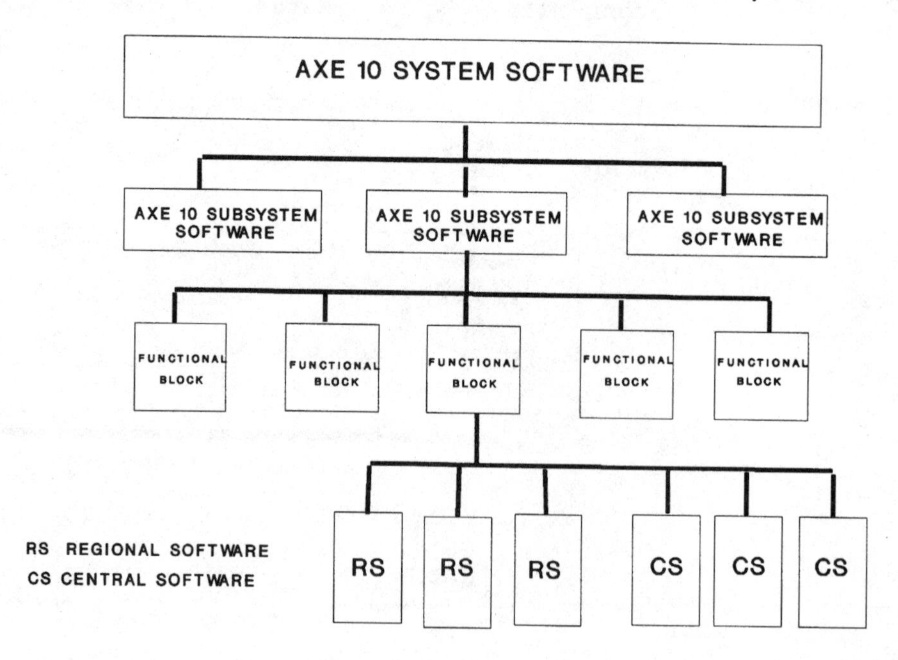

Figure 10-6 Ericcson AXE 10 software architecture

- Functional block: blocks of software that make up a subsystem
- Regional software: software that is common to RPS and makeup functional blocks
- Central software: software common to CPS and makeup functional blocks

A Simple Line-to-Line Call through the Switch Customer A calls Customer B in the same switch. When customer A goes off-hook, it is detected by the line switch module, which is interconnected to the SSS. The SSS signals the RP with an off-hook condition which in turn asks the CP for a time slot. The CP determines the line status and orders RPS to attach the digital receiver. The CP analyzes the digits and, if they are valid, sends a signal to the RP to ring customer B's phone. When customer B answers, the CP sends the required signals to the RP and associated GSS for a talking path connection between customer A and customer B. When either customer hangs up, the line switch module detects the on-hook condition and the path is idled.

10.1.4 Siemens Stromberg Carlson's EWSD

System Overview The EWSD* digital switching system is a class 5 digital switching system. The EWSD switch provides POTS and business services and also supports advanced services such as ISDN, STP, SCP, and AIN. It was first put in service in 1980.

Its architecture can be classified as quasi-distributed since it still maintains control of the system via a coordination processor. Its switching fabric is time-multiplexed and is classified as a time-space-time (TST) network.

Hardware Architecture[7] The EWSD digital switching system has a modular design, and its hardware architecture is shown in Fig. 10.7. The basic subsystems of the EWSD switch are

- Switching network
- Line trunk group
- Common signaling network control

*EWSD is a registered trademark of Siemens AG.

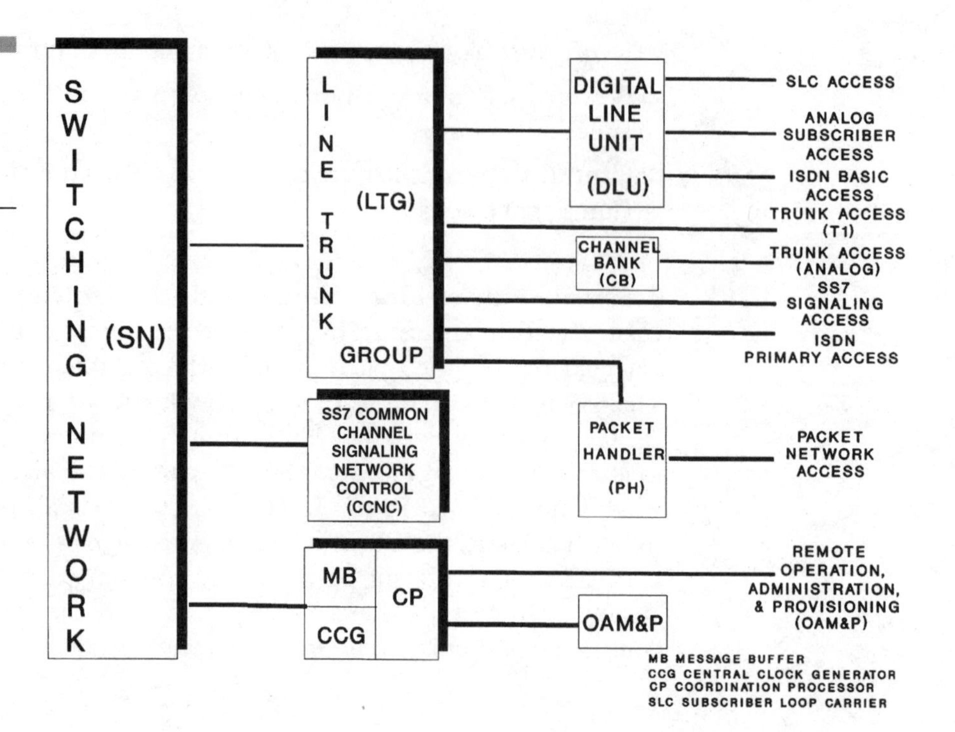

Figure 10.7 Siemens Stromberg-Carlson EWSD switch hardware architecture

- Message buffer, central clock generator, and coordination processor
- Digital line unit
- Channel bank
- Packet handler
- Operation, administration, and provisioning
- Remote control unit (not shown)

Switching Network (SN). This is the switching fabric of the EWSD switch. It connects lines, trunks, and signaling connections between subsystems via the line trunk group interface. The switching network is based on a time-space-time (TST) configuration.

Line Trunk Group (LTG). This interface multiplexes and controls traffic between line and trunk interfaces and the switching network.

Common Channel Signaling Network Control (CCNC). This interface supports SS7 signaling between the EWSD switch and SSPs, STPs, and SCPs.

Message Buffer (MB), Central Clock Generator (CCG), and Coordination Processor (CP). The message buffer (MB) processes signals between the network and the coordination processor. The central clock generator (CCG) provides clock signals for the EWSD switch. The coordination processor coordinates the activities of other system processors.

Digital Line Unit (DLU). The digital line unit interconnects analog subscribers, ISDN basic access, and PBX lines to the LTG and also supports SLC access. This interface provides line concentration, too.

Channel Bank (CB). This interface connects analog trunks to the LTG.

Packet Handler. The packet handler supports ISDN packet subscribers.

Operation, Administration, and Provisioning (OAM&P). The operation, administration, and provisioning subsystem supports the human-machine interface, system diagnostics, fault handling, traffic, and other measurements.

Remote Control Unit (RCU). It supports remote operation of DLUs and stand-alone local service.

Software Architecture[8] A basic architecture for the EWSD software is shown in Fig. 10.8. The structure of the EWSD software is highly modular. The EWSD software can broadly be classified into OAM&P software, exchange software, support software, and customer premises software. As the names imply, OAM&P software supports all operational, administration, and provisioning activities. The exchange software supports call processing, the coordination processor, and all other peripheral processors. The support software helps to engineer and plan the EWSD switch. The customer premises software supports activities of centrex attends. All peripheral units such as the MB, DLU, LTG, and CCNC are loaded with specific software for their respective functionalities. The switching network is also loaded with its specific software.

A Simple Call through the Switch A very simplified explanation of a simple line-to-line call through an EWSD switch follows:

Customer A calls customer B in the same switch. When customer A goes off-hook to call customer B, the call origination request is detected by the subscriber line module digital line unit (DLU). It sends a message to the

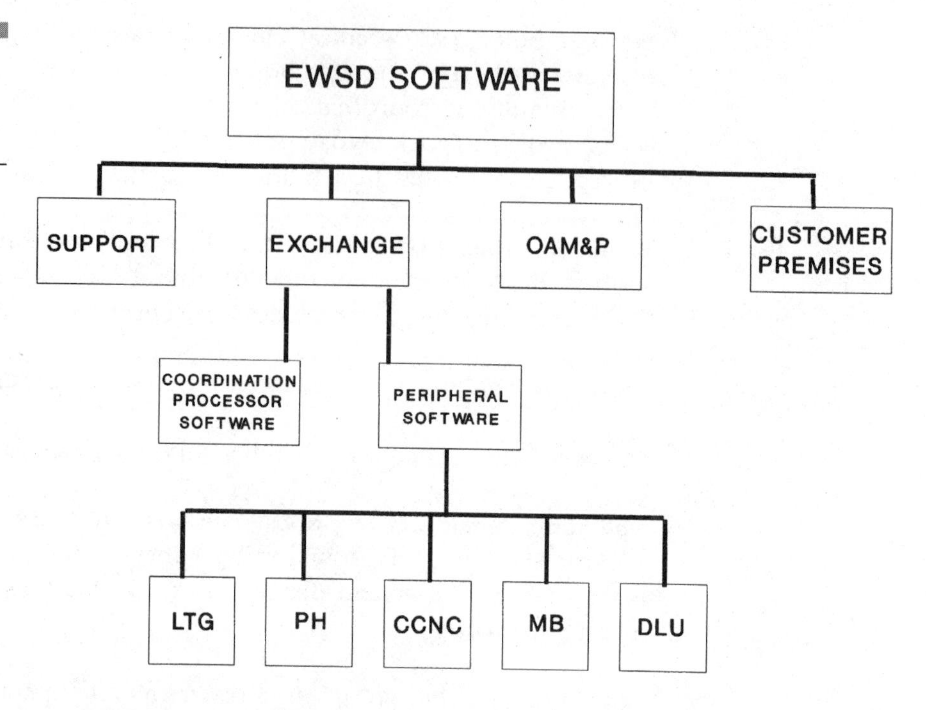

Figure 10.8 Siemens Stromberg-Carlson EWSD software architecture

CP via the SN and validates customer A's line. A digit receiver is attached to the line, and a dial tone is provided to customer A by the LTG via the subscriber line module. After the first digit is dialed and received by the line module, the dial tone is removed from customer A's telephone. The dialed digits are passed to the CP for analysis. If the dialed number is valid, time slots are assigned for customer A and customer B. If the dialed number is incorrect, for instance, has a wrong prefix, an announcement or a tone is given to customer A. Customer B's line is checked for busy/idle status, and a power ringing is applied to customer B's line if it is found to be idle. An audible ringing is simultaneously applied to customer A's line. When customer B answers, a cut-through path is provided through the network via previously assigned time slots. If either customer disconnects, the line module detects the on-hook condition and idles the connection.

10.1.5 GTE's GTD-5 EAX Switch—Electronic Automatic Exchange

System Overview The GTD-5 EAX* switch is a medium-size to large local and toll switch. The GTD-5 EAX switch provides POTS and centrex services and also supports advanced services such as ISDN, STP, SCP, and AIN. It was first put in service in 1982.

Its architecture can be classified as quasi-distributed since it still maintains control of the system via an administrative processor. Its switching fabric is time-multiplexed and is classified as a time-space-time (TST) network.

Hardware Architecture[9] The GTD-5 EAX switch has a modular design, and its hardware architecture is shown in Fig. 10.9. The basic subsystems of the GTD-EAX switch are:

- Telephone processor complex
- Common memory
- Space interface controller
- Administrative processor complex
- Message distributor circuit
- Time switch and peripheral control unit

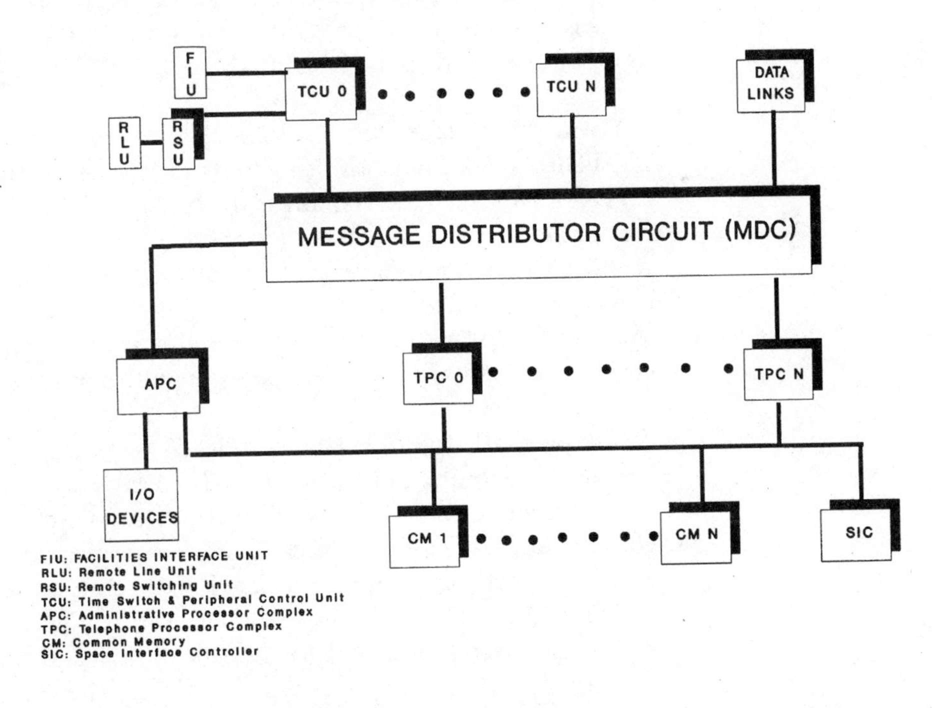

Figure 10.9 GTE GTD-5 EAX switch hardware architecture

[9] GTD-5 EAX is a trademark of GTE Communication Systems.

- Remote line unit
- Remote switching unit

Lines and trunks are connected through the facilities interface units (FIUs) to the time switch and peripheral control units. Remote lines are connected through the remote line units to the remote switching units (RSUs). The RSUs and TSUs perform similar functions; however, if the remote unit becomes isolated from the base unit, the RSUs can provide stand-alone telephony service to remote subscribers. The TCUs control the FIUs and provide most basic call processing functions. The switching fabric for the GTD-5 EAX switch is a space switch, which is connected to each TCU and controlled by the space interface controller (SIC). Time-switched input is provided to the space switch through the time slots assigned by the TCUs. Calls are switched in time-space-time (TST) mode. The interprocessor communication in the switch is provided by the message distributor circuit (MDC). The telephone processor complex (TPC) provides high-level call processing functions such as call routing and space switch control. It also keeps track of all calls in common memories (CMs) that are accessible to all processors. The administrative processor complex (APC) provides access to all input/output devices such as tapes, disks, and control terminals.

Software Architecture[10] A high-level software architecture is shown in Fig. 10.10. All administrative functions for operating and maintaining the GTD-5 EAX switch are contained in the administrator processor software that resides in the APC. All the telephony functions software is installed in the TPCs. It performs all high-level call-related functions such as network control and control of devices required during a call. All basic telephonic control software for scanning lines, line control, trunk control, service circuits, etc. resides in the TCU and for remote operations in the RCU.

A Simple Call through the Switch A very simplified explanation of a simple line-to-line call through a GTD-5 EAX switch follows:

Customer A calls customer B in the same switch. When customer A goes off-hook to call customer B, the call origination request is detected by the facilities interface unit (FIU). It sends a message to the TCU which contacts the TPC via the MDC and validates customer A's line. A digit receiver is attached to the line, and a dial tone is provided to customer A. After the first digit is dialed, the dial tone is removed from customer A's phone and the FIU forwards the dialed digits to

Figure 10.10
GTE GTD-5 EAX software architecture

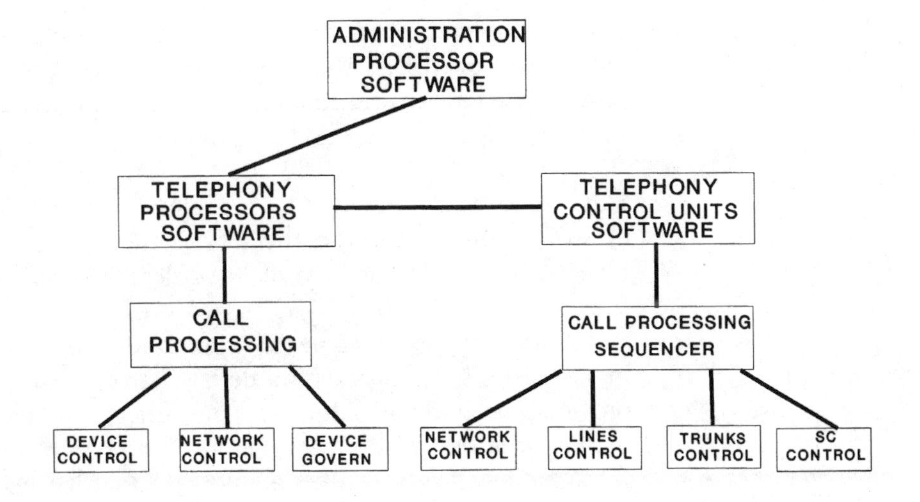

the TCU. The dialed digits are passed to TPC for analysis. If the dialed number is valid, then time slots are assigned for customer A and customer B and the path is recorded in the common memory. If the dialed number is incorrect, an announcement or a tone is given to customer A. Customer B's line is checked for busy/idle status, and a power ringing is applied to customer B's line if it is found to be idle. An audible ringing is simultaneously applied to customer A's line. When customer B answers, a cut-through path is provided through the network via previously assigned time slots. If either customer disconnects, the FIU detects the on-hook condition and informs the TCU. TCU sends a message through the MDC to TPCs and SIC to idle the connection.

10.1.6 NEC's NEAX 61

System Overview The NEAX 61* digital switching system is a medium to large class 5 digital switching system. It switches local and toll services. It also supports advanced services such as ISDN, STP, SCP, and AIN. It was first put in service in 1979.

*NEAX 61 is a trademark of NEC Corporation.

Its architecture can be classified as quasi-distributed since it still maintains control of the system via an operations and maintenance processor. Its switching fabric is time-multiplexed and is classified as a time-space-space-time (TSST) network.

Hardware Architecture[11] The lines and trunks interconnect to the network module through the service interface modules. See Fig. 10.11. Line modules (LMs) and trunk modules (TMs) reside in the application subsystem (not shown). The network module provides concentration and a time-space-space-time (TSST) path for the calls through the switching system. The network module is controlled by call processors which provide practically all call processing functions. All call processing information is stored in the local memory and common memory which are accessible to all other call processors (CLPs). The operation and maintenance processor (OMP) provides all system maintenance functions and supports operation of other CLPs. All inputs and outputs to the switch are provided by the OMP.

Software Architecture[12] The software of the NEAX 61 switch can be classified into two categories: the operating system (OS) and the application system (APL). The OS includes the executive control program (EP), diagnostic program (DP), and the fault processing (FP) program. The APL includes the call processing programs (CP) and the administration program (AP). The administration programs supports all input/output functions of the switch and office data. See Fig. 10.12.

A Simple Call through the Switch A very simplified explanation of a simple line-to-line call through an NEAX 61 switch follows:

Customer A calls customer B in the same switch. When customer A goes off-hook to call customer B, the call origination request is detected by the line interface module. It sends a message to associated call processor (CLP) through the network module. The network module switches call connection in time-space-space-time mode. The CLP validates customer A's line. A digit receiver is attached to the line, and a dial tone is provided to customer A. When the first dialed number is received by the LM, the dial tone is removed from customer A's receiver. The dialed digits are passed to the CLP for analysis. If the dialed number is valid, then time slots are assigned for customer A and customer B and the call information is registered in the local and common memory. If the dialed number is incorrect, an announcement or

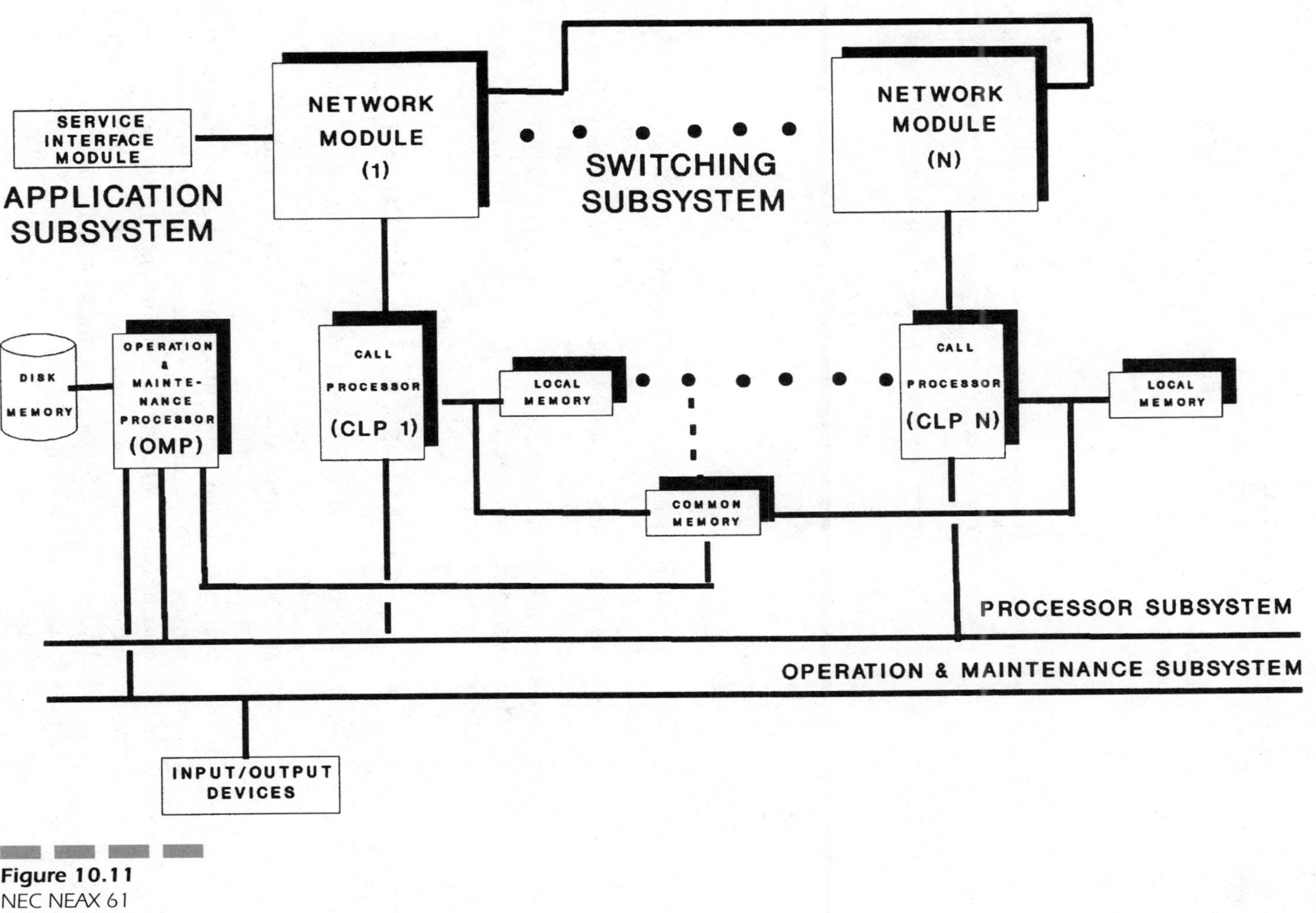

Figure 10.11
NEC NEAX 61 switch hardware architecture

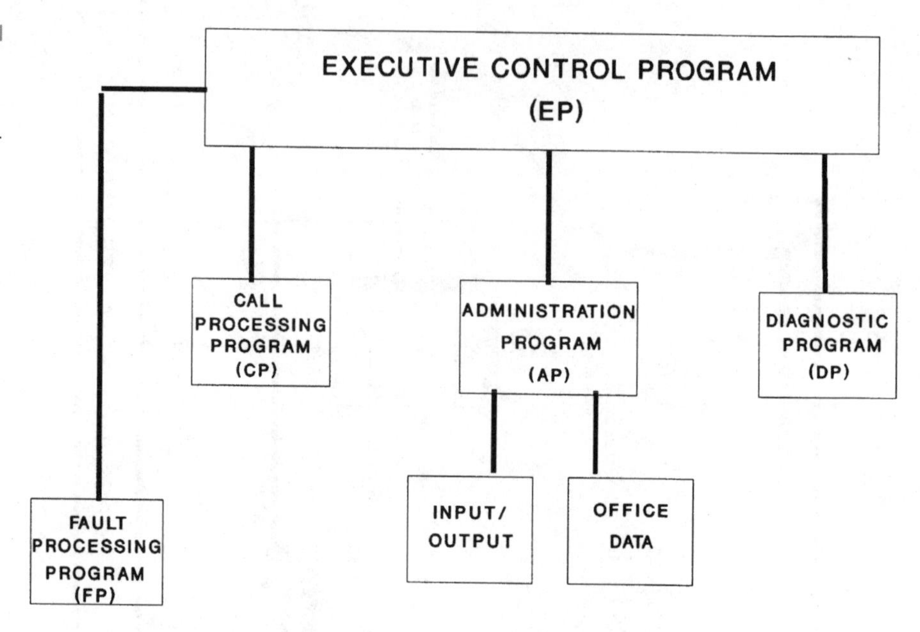

Figure 10.12 NEC NEAX 61 software architecture

a tone is given to customer A. Customer B's line is checked for busy/idle status, and a power ringing is applied to customer B's line if it is found to be idle. An audible ringing is simultaneously applied to customer A's line. When customer B answers, a cut-through path is provided through the network via previously assigned time slots. If either customer disconnects, the LM detects the on-hook condition and idles the connection.

10.1.7 Alcatel's System 12* Digital Exchange

System Overview The System 12 digital exchange is a class 5 digital switching system. The System 12 switch provides POTS and centrex services and supports advanced services such as ISDN and AIN. It was first put in service in 1978.

The System 12 switch has a fully distributed architecture. It does not have a central control processor, and each module can communicate with and connect to other modules through its digital switching network. Network messages are used for all communications within the switch. Its switching fabric is time-multiplexed and switches through multiple time-space (TS) stages.

* System 12 is a registered trademark of Alcatel.

Hardware Architecture[13] The System 12 digital switching system has a highly modular design, and its hardware architecture is shown in Fig. 10.13. The basic subsystems, called *terminal control elements* (*TCEs*) for the System 12 switch, are

- Data interface
- Analog subscriber
- Digital subscriber
- Remote interface
- Common channel
- Service circuits
- Operator interface
- Analog trunk
- Digital trunk
- Exchange interface
- Computer peripherals
- Clock and tones

The basic functions of the above subsystem are evident from their names, and they are not duplicated, except the clock and tones module. It is duplicated for reliability. The System 12 switch can be expanded by

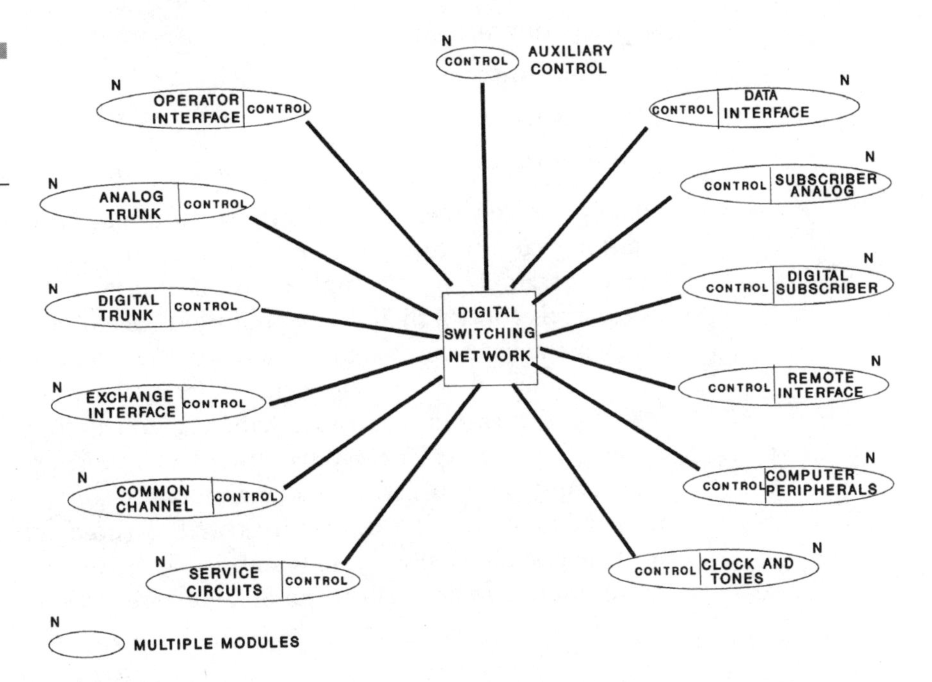

Figure 10-13 Alcatel System 12 switch hardware architecture

adding TCEs and is represented by N in Fig. 10.13. The other important component of the System 12 switch is the auxiliary control element (ACE). Each TCE is assigned to an ACE. The ACE unit keeps all information about the TCEs assigned to it and performs call control functions. The other function of the ACE is to allocate exchange resources and perform digit analysis.

Digital Switching Network. The digital switching network (DSN) of the System 12 switch consists of an access switch for the first stage of switching and group switches for the second, third, and fourth stages of switching. The group switches have multiple switching planes and can be grown by adding planes. The terminal control elements access the DSN via messages. These messages allow the TCE to hunt for an available path through the DSN for connection to other terminal control elements. The time slots are reserved through the DSN. A return path through the DSN is also hunted and is not the same as the forward path. The connection process between the TCEs in the DSN is similar to step-by-step switching.

Software Architecture[14] A high-level structure of System 12 software is shown in Fig. 10-14. It has the following functional areas:

- Operating system and database
- Telephony support
- Call handling
- Maintenance
- Administration

The operating system controls the execution of application programs. It supports communications between modules, scheduling, timing, etc. Since System 12 uses a highly distributed architecture, most of the peripherals control their own applications. Since each module maintains its own database, the system keeps these databases up to date.

The function of the telephony support software is to manage hardware devices through software handlers, and it supports signaling and customer charging. The call-handling software supports call setup and release and call translation. The maintenance software supports centralized maintenance of the switch as well as those actions that are done routinely such as system audits and diagnostic tests. The centralized maintenance system coordinates diagnostic tests and generates test

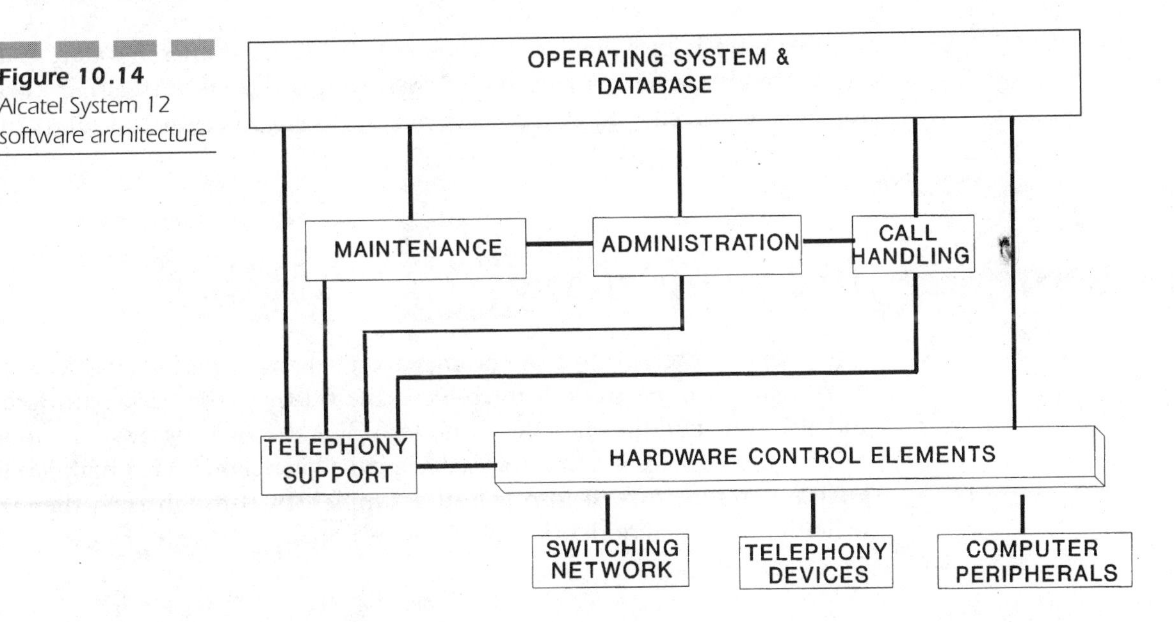

Figure 10.14 Alcatel System 12 software architecture

results. The administration software helps the maintenance personnel to operate the switching system. It supports in adding or deleting of lines and trunks. It also supports plant measurement.

A Simple Line-to-Line Call through the Switch A very simplified explanation of a simple line-to-line call through a System 12 switch follows:

Customer A calls customer B in the same switch. When customer A goes off-hook to call customer B, the call origination request is detected by the line analog subscriber module or terminal control element (TCE) for subscriber A. It sends a message to an associated auxiliary control element (ACE) through the digital switching network which validates customer A's line. A digits receiver is attached to the line, and a dial tone is provided to customer A by a connection to a service circuit unit through the DSN. The dialed number is received by the service circuit and is sent to the ACE for analysis. The dial tone is removed from customer A's receiver after the first digit is dialed. If the dialed number is valid, customer B is identified and customer B's equipment number is forwarded to the ACE of customer A. If the dialed number is incorrect, an announcement or a tone is given to customer A. The TCE of customer A initiates a connection request to the TCE of customer B. Customer B's line is checked by the TCE of customer B, and a path is established through the DSN. Ringing is provided to the called customer B by customer B's TCE, and an audible tone is provided to customer A. When customer B answers, a cut-through speech path is provided through the DSN. If either customer disconnects, the TCE detects the on-hook condition and idles the connection via the ACEs.

10.2 Summary

This chapter covered most of the major digital switching systems in the North American network. It introduced the reader to the basic hardware and software building blocks of modern digital switching systems, and covered the switching fabric for each digital switch along with high-level software architecture. It also traced a simple call through each digital switching system and identified basic steps necessary for call processing.

REFERENCES

1. D. L. Carney, J. I. Cochrane, L. J. Gitten, E. M. Prell, and R. R. Staehler, "The 5ESS Switching System: Architectural Overview," *AT&T Technical Journal*, vol. 64, no. 6, July-August 1985, pp. 1339–1356.
2. J. P. Delatore, R. J. Frank, H. Oearing, and L. C. Stecher, "The 5ESS Switching System: Operational Software," *AT&T Technical Journal*, vol. 64, no. 6, July-August 1985, pp. 1357–1383.
3. Northern Telecom, Integrated Marketing Communications, *DMS-100 Advantage-Technology, Services and Support*, publication 50157.11/01-93, issue 1, January 1993.
4. Northern Telecom, Integrated Marketing Communications, *DMS SuperNode Software*, publication 50039.17/07-93, issue 1, July 1993.
5. M. Eklund, C. Larson, and K. Sörme, "AXE 10—System Description," *Ericsson Review*, vol. 53, no. 2, 1976, pp. 70–89.
6. B. Eklundh and David Rapp, "Load Study of the AXE 10 Control System," *Ericsson Review*, vol. 59, no. 4, 1982, pp. 208–216.
7. *Siemens Stromberg-Carlson EWSD Digital Switching System: System Description*, Document EWSD PD-10/95, pp. 4–9.

8. Ibid., pp. 28–30.
9. M. Wienshienk and S. Magnusson, "Low Capacity Switching Units in the GTD 5 EAX System," *Proceedings of International Switching Symposium 1984,* vol. 3, Session 33 B, paper 4, 1985, pp. 1–5.
10. J. Brown and T. Lenz, "Transportability of Software," *Proceedings of International Switching Symposium 1984,* vol. 3, Session 34 A, paper 5, 1984, pp. 1–7.
11. Rajiv Jaluria, Yoshio Nishimura, and Robert E. McFadden, "Digital Switching System for Specialized Common Carrier," *Proceedings of International Switching Symposium 1984,* vol. 3, Session 34 B, paper 7, 1984, pp. 3–4.
12. Ibid., p. 6.
13. J. Das, K. Strunk, and F. Verstraete, "ITT 1240 Digital Exchange Hardware Description," *Electrical Communication—The Technical Journal of ITT,* vol. 56, no. 2/3, 1981, pp. 135–147.
14. L. Katzschner and F. Van den Brand, "ITT 1240 Digital Exchange Software Concepts and Implementation," *Electrical Communication—The Technical Journal of ITT,* vol. 56, no. 2/3, 1981, pp. 173–183.

INDEX

D

O

P

R

S

T

About the Author

Syed R. Ali is currently a Senior Engineer in the Integration and Implementation organization at Bellcore (Piscataway, NJ) and has over 27 years' experience in telephone engineering and digital switching system analysis. He received a BSEE from East Pakistan University of Engineering and Technology, Dhaka, in 1967; a MSEE degree from Tuskegee Institute, Tuskegee, Alabama, in 1971; and an advanced degree in electrical engineering from New Jersey Institute of Technology, Newark, New Jersey, in 1982.

At Bellcore he pioneered the methodologies for analyzing root causes for switching system downtimes. He is the author of many Bellcore's generic requirements in functional areas such as software metrics (IPQM), switching system testing, software patching, firmware, and object-oriented process metrics (OOPM).

Syed Ali is a frequent speaker at many telecommunications conferences and the author of numerous papers on telecommunications system reliability and maintainability.

The author is a past chairman of the IEEE Communications Society and co-founder and chairman of IEEE Computer Society, New York Section. He is a senior member of IEEE as well as a member of ACM and the National Honor Society for Electrical Engineers, Eta Kappa Nu.